The Cambridge Star Atlas covers the entire sky, both northern and southern latitudes, in an attractive format that is suitable for beginners and experienced astronomical observers. There is a series of monthly sky charts, followed by an atlas of the whole sky, arranged in 20 overlapping full-color charts. Each chart shows stars down to magnitude 6.5, together with about 900 non-stellar objects, such as clusters and galaxies, which can be seen with binoculars or a small telescope. There is a comprehensive map of the Moon's surface, showing craters and other named features.

Wil Tirion is the world's foremost designer of astronomical maps. For this edition he has devised improved versions of all the charts, and the text and star data have been completely revised based on the latest information. Clear, authoritative and easy to use, *The Cambridge Star Atlas* is an ideal reference atlas for sky watchers everywhere.

THE CAMBRIDGE

STAR

ATLAS

THIRD EDITION

WIL TIRION

CAMBRIDGE
UNIVERSITY PRESS

PUBLISHED BY THE PRESS SYNDICATE OF THE UNIVERSITY OF CAMBRIDGE
The Pitt Building, Trumpington Street, Cambridge, United Kingdom

CAMBRIDGE UNIVERSITY PRESS
The Edinburgh Building, Cambridge CB2 2RU, UK
40 West 20th Street, New York, NY 10011-4211, USA
477 Williamstown Road, Port Melbourne, VIC 3207, Australia
Ruiz de Alarcón 13, 28014 Madrid, Spain
Dock House, The Waterfront, Cape Town 8001, South Africa

http://www.cambridge.org

First published 1991
Second edition 1996
Reprinted 1998, 1999
Third edition 2001
Reprinted 2004

Printed in the United Kingdom at the University Press, Cambridge

Typeface Adobe Caslon 11/14pt   *System* QuarkXpress® [wv]

A catalogue record for this book is available from the British Library

Library of Congress Cataloguing in publication data

Tirion, Wil.
    Cambridge star atlas / Wil Tirion. – 3rd ed.
        p.   cm.
    Includes bibliographical references.
    ISBN 0 521 80084 6
    1. Stars – Atlases.   I. Title.
QB65.T537 2000
523.8'022'3–dc21   00–059909

ISBN 0 521 80084 6 hardback

CONTENTS

PREFACE

Anyone who looks up at the starry sky at night and wonders how to find a way among all those stars will need some kind of sky-guide or atlas, but very different needs must be met. The casual stargazer will first want to learn what can be seen with the unaided eye; the names of the stars, the constellations and where or when to look for Orion, the Great Bear, or Andromeda. The more advanced observer, with access to a good pair of binoculars or a small telescope, wants to know more: where is the Whirlpool Galaxy, where the North-America Nebula, or the globular cluster M13?

The Cambridge Star Atlas offers help for both. It includes a series of twenty-four monthly sky maps, designed to be of use for almost anywhere on Earth and a series of twenty detailed star charts, covering the whole heavens, with all stars visible to the naked eye under good circumstances. These twenty star charts also show a wealth of star clusters, nebulae and galaxies. Some of these can be seen without optical help, but for most a small or average-size telescope is needed. Accompanying the charts are tables, offering accurate positions and more details of these objects, as well as information about interesting double and variable stars. For this third edition of *The Cambridge Star Atlas* a complete new series of charts has been created and, consequently, the chart's tables have been revised as well.

At the end of the atlas you will find a series of six all-sky maps, showing the sky in a special projection, centered on the Galactic Equator. They show the distribution of stars, open and globular clusters, planetary and diffuse nebulae and galaxies, in relation to our own Milky Way.

Probably the most satisfying sight for a beginner, using a pair of binoculars or a small telescope, is the Moon. Therefore, *The Cambridge Star Atlas* starts with a not-too-complicated Moon map, showing the most important features on its surface. Since the Moon is our neighbor in space, and because it is usually the first thing we notice in the night sky, the Moon map is placed in front, where it belongs.

Happy stargazing!

Wil Tirion

THE MOON

The Moon is, apart from the Sun, the brightest object in the sky. Although the Sun and the Moon appear almost equal in size, they are quite different. The Sun is the central body of our Solar System, and all planets, including the Earth, orbit around it. The Sun measures 1.4 million kilometers across, and is at a distance of roughly 150,000,000 kilometers. The Moon is much smaller, 'only' 3,476 kilometers across, approximately one quarter of the Earth's diameter, and at an average distance of 384,400 kilometers. It orbits, not the Sun, but our own planet, in a little more than 27 days.

Although we often refer to the Moon as 'shining' it does not of itself give any light. It only reflects the light it receives from the Sun. This is the reason why the appearance of the Moon changes as it orbits the Earth. This aspect of the Moon, sometimes visible as a thin crescent in the western sky, after sunset, and sometimes as a full disk, lightening up the middle of the night, is confusing to most people. The reason

for this can be best explained in a diagram (Figure 1).

The illustration is not drawn to scale, but shows you what happens. The Earth is at the center and the Moon's orbit is drawn as a circle. During its orbit around the Earth we see a different portion of the illuminated side of the Moon's surface. When the Moon is approximately between the Sun and the Earth, we see only the dark side of it. We call this *new Moon*. The Moon is not visible at all. After a few days we see a small crescent in the evening sky; a part of the illuminated side is peeping around the edge. After almost a week, half of the disk is lit and we call this *first quarter*. Another week later we see the complete disk. This is *full Moon*. Next comes the *last quarter*, and then back to *new Moon* again. From one new Moon to the next takes about 29.5 days, fully two days longer than it takes the Moon to orbit the Earth. The reason for this is that, in the time the Moon revolves around the Earth, the Earth also moves in its orbit around the Sun.

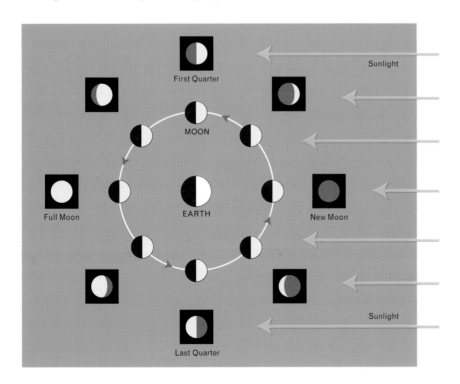

Figure 1

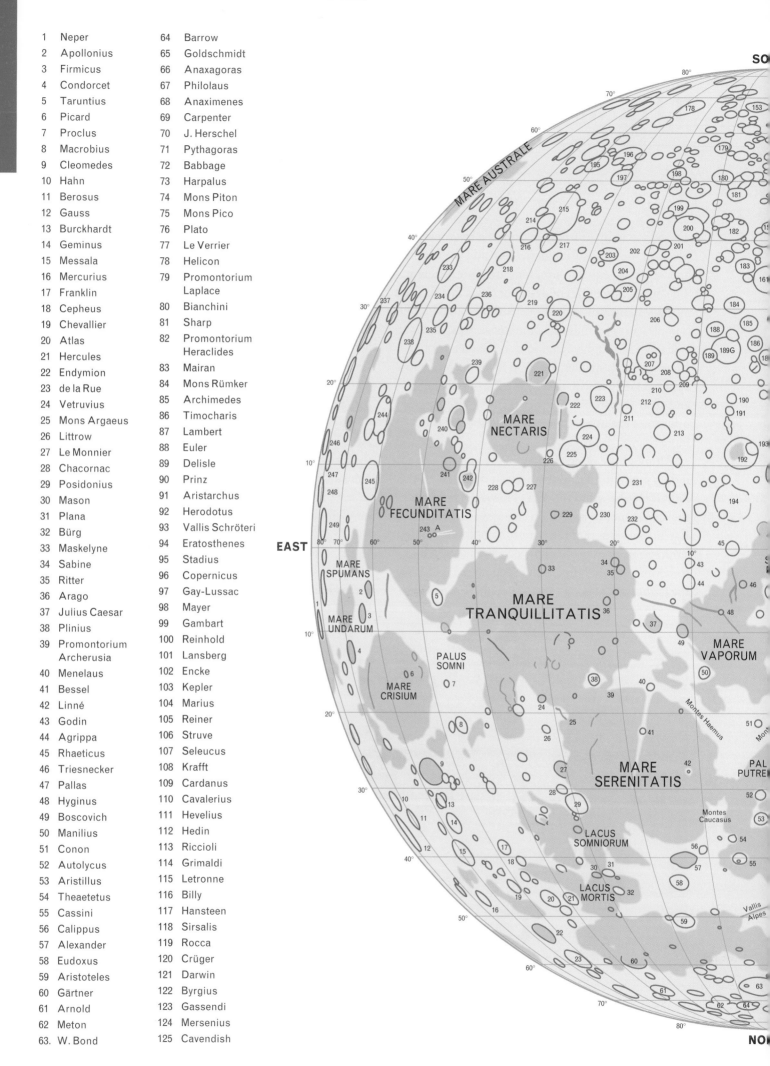

1	Neper	64	Barrow
2	Apollonius	65	Goldschmidt
3	Firmicus	66	Anaxagoras
4	Condorcet	67	Philolaus
5	Taruntius	68	Anaximenes
6	Picard	69	Carpenter
7	Proclus	70	J. Herschel
8	Macrobius	71	Pythagoras
9	Cleomedes	72	Babbage
10	Hahn	73	Harpalus
11	Berosus	74	Mons Piton
12	Gauss	75	Mons Pico
13	Burckhardt	76	Plato
14	Geminus	77	Le Verrier
15	Messala	78	Helicon
16	Mercurius	79	Promontorium Laplace
17	Franklin	80	Bianchini
18	Cepheus	81	Sharp
19	Chevallier	82	Promontorium Heraclides
20	Atlas	83	Mairan
21	Hercules	84	Mons Rümker
22	Endymion	85	Archimedes
23	de la Rue	86	Timocharis
24	Vetruvius	87	Lambert
25	Mons Argaeus	88	Euler
26	Littrow	89	Delisle
27	Le Monnier	90	Prinz
28	Chacornac	91	Aristarchus
29	Posidonius	92	Herodotus
30	Mason	93	Vallis Schröteri
31	Plana	94	Eratosthenes
32	Bürg	95	Stadius
33	Maskelyne	96	Copernicus
34	Sabine	97	Gay-Lussac
35	Ritter	98	Mayer
36	Arago	99	Gambart
37	Julius Caesar	100	Reinhold
38	Plinius	101	Lansberg
39	Promontorium Archerusia	102	Encke
40	Menelaus	103	Kepler
41	Bessel	104	Marius
42	Linné	105	Reiner
43	Godin	106	Struve
44	Agrippa	107	Seleucus
45	Rhaeticus	108	Krafft
46	Triesnecker	109	Cardanus
47	Pallas	110	Cavalerius
48	Hyginus	111	Hevelius
49	Boscovich	112	Hedin
50	Manilius	113	Riccioli
51	Conon	114	Grimaldi
52	Autolycus	115	Letronne
53	Aristillus	116	Billy
54	Theaetetus	117	Hansteen
55	Cassini	118	Sirsalis
56	Calippus	119	Rocca
57	Alexander	120	Crüger
58	Eudoxus	121	Darwin
59	Aristoteles	122	Byrgius
60	Gärtner	123	Gassendi
61	Arnold	124	Mersenius
62	Meton	125	Cavendish
63.	W. Bond		

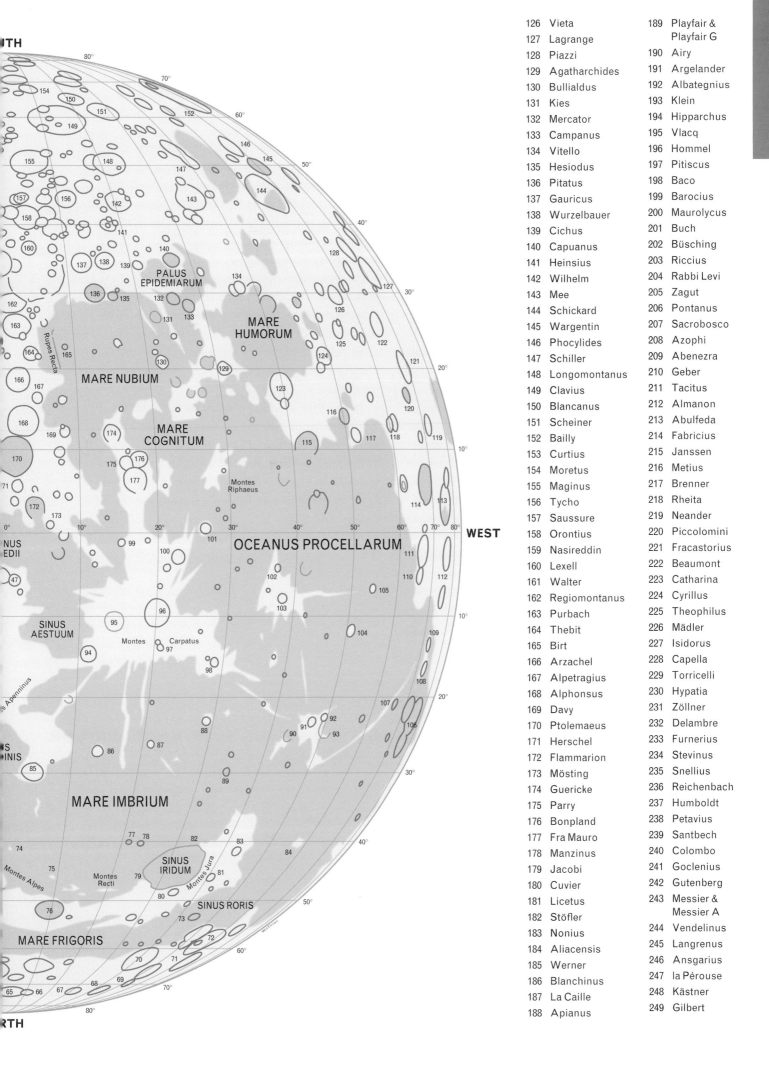

126	Vieta	189	Playfair & Playfair G
127	Lagrange	190	Airy
128	Piazzi	191	Argelander
129	Agatharchides	192	Albategnius
130	Bullialdus	193	Klein
131	Kies	194	Hipparchus
132	Mercator	195	Vlacq
133	Campanus	196	Hommel
134	Vitello	197	Pitiscus
135	Hesiodus	198	Baco
136	Pitatus	199	Barocius
137	Gauricus	200	Maurolycus
138	Wurzelbauer	201	Buch
139	Cichus	202	Büsching
140	Capuanus	203	Riccius
141	Heinsius	204	Rabbi Levi
142	Wilhelm	205	Zagut
143	Mee	206	Pontanus
144	Schickard	207	Sacrobosco
145	Wargentin	208	Azophi
146	Phocylides	209	Abenezra
147	Schiller	210	Geber
148	Longomontanus	211	Tacitus
149	Clavius	212	Almanon
150	Blancanus	213	Abulfeda
151	Scheiner	214	Fabricius
152	Bailly	215	Janssen
153	Curtius	216	Metius
154	Moretus	217	Brenner
155	Maginus	218	Rheita
156	Tycho	219	Neander
157	Saussure	220	Piccolomini
158	Orontius	221	Fracastorius
159	Nasireddin	222	Beaumont
160	Lexell	223	Catharina
161	Walter	224	Cyrillus
162	Regiomontanus	225	Theophilus
163	Purbach	226	Mädler
164	Thebit	227	Isidorus
165	Birt	228	Capella
166	Arzachel	229	Torricelli
167	Alpetragius	230	Hypatia
168	Alphonsus	231	Zöllner
169	Davy	232	Delambre
170	Ptolemaeus	233	Furnerius
171	Herschel	234	Stevinus
172	Flammarion	235	Snellius
173	Mösting	236	Reichenbach
174	Guericke	237	Humboldt
175	Parry	238	Petavius
176	Bonpland	239	Santbech
177	Fra Mauro	240	Colombo
178	Manzinus	241	Goclenius
179	Jacobi	242	Gutenberg
180	Cuvier	243	Messier & Messier A
181	Licetus	244	Vendelinus
182	Stöfler	245	Langrenus
183	Nonius	246	Ansgarius
184	Aliacensis	247	la Pérouse
185	Werner	248	Kästner
186	Blanchinus	249	Gilbert
187	La Caille		
188	Apianus		

Observing the Moon

Even with a simple pair of binoculars you can see interesting features on the Moon's surface, and a small telescope will reveal even more details of our neighbor in space. The best time to watch the Moon is not when it is full, but rather around its first or last quarter. Then the Moon is illuminated by the Sun from one side and especially near the terminator, the line dividing the lit and the unlit halves of the Moon, there is strong relief, because the surface is illuminated from a very low angle, resulting in long shadows. At full Moon you do not see any relief, since you are then looking from approximately the same direction as the Sun's rays come from. But full Moon is an ideal time to study the differences between the dark and light areas of the surface.

On the Moon map (on the previous two pages), all craters and crater-like features are outlined in dark grey-green, while the larger *maria* are shaded in a much lighter grey-green. *Maria* is the plural form of the Latin word *mare,* meaning sea, and the name was given by the first observers who believed that these dark areas on the Moon really were seas and oceans. Although we now know there is no water on the Moon, the name persists, as also do the names *lacus* (lake) and *oceanus* (ocean). All *maria* and mountain ranges are labeled on the map itself, while the craters are numbered (to avoid overloading). You can find the names in the columns to the left and the right of the map.

Most craters on the Moon are believed to be the result of the impact of meteors: pieces of rock and metal from space. Our Earth is well protected against the impact of meteors by the atmosphere, which causes meteors to burn and evaporate. Only the larger ones reach the surface; we call these meteorites. But the Moon does not have an atmosphere, so every meteor captured by the Moon's gravity will crash into the surface.

Because the Moon rotates 360° on its axis in exactly the same time that it takes to complete one orbit around the Earth, we always see the same side of the Moon. However the Moon's orbit is inclined about 5° to the *ecliptic* (see chapter *The star charts,* page 32), making it move slightly above and below the plane of the Earth's orbit around the Sun, and the Moon's axis is also tilted about 1.5°. The combined result is that we can also look about 6.5° 'over' the North and South Pole of the Moon. Moreover, since the Moon's orbit is not really a circle, but an ellipse, it does not move at a constant speed, though its rotation speed remains the same. Thus it moves a little from left to right, as if it were shaking its head very slowly. Therefore, we can look around the edges, by up to 7°. Sometimes *Mare Crisium* (in the northeastern quadrant of the Moon) appears very close to the edge, and sometimes it is closer to the center, and has a less elliptical appearance. The elliptical appearance of Mare Crisium, as well as that of craters close to the edge, is of course caused by perspective.

In most telescopes the image of the object that is viewed is upside down. For that reason the Moon map is placed with south on the top. So, if you are observing the Moon through a pair of binoculars, you will have to turn the atlas around. Unless you are living in the southern hemisphere; then it is the other way round!

THE MONTHLY SKY MAPS

Looking up at the night sky you will probably see an unorderly scattered arrangement of stars, but anyone who looks up more frequently will start to see that the stars do not change places in relation to each other. Long ago, people began giving names to these fixed groupings of stars, and today we refer to these groups as 'constellations'. Most people recognize one or two constellations; for example, they probably know what the Great Bear looks like. But why is it always in a different position in the sky? And why is it not possible to find Orion during a night in June or July? This changing aspect of the sky is often confusing to the casual stargazer. So, the first thing you have to learn is how the sky moves.

It is important to know that the star patterns themselves do not change, at least not in a single human life span. It is only over a period of centuries that the positions of some neighboring stars change in a way that can be detected with the unaided eye. All these groupings of stars and constellations can be regarded as being fixed to a huge imaginary sphere, with the Earth placed in the center. No matter where on Earth we are, we can always see just one half of this sphere. So it is not hard to understand that, when we move to another part of the Earth, the visible part of the sphere also changes. If we stand at the North Pole we will only see the northern half of the heavens, while at the South Pole we will see only the southern half. But it is not all that simple. Two more factors affect the appearance of the sky. First there is the daily rotation of the Earth around its axis, which causes the Sun to come up in the east and set in the west. The same thing happens with the other objects in the sky. In fact, it looks as if the entire heavenly sphere rotates around an axis that is an extension of the Earth's axis. Apart from that there is the orbital movement of the Earth, which makes the appearance of the sky change over the seasons. If you look at one constellation, let us say Orion, at midnight on the first day of January and take note of its position, and then look every

successive night at about the same time, you will notice that the stars reach the same position a few minutes earlier each night. One month later, on the first day of February, Orion will already be in that position at 10 PM. The appearance of the night sky thus changes slowly, until one year later the Earth has reached the same point in its orbit again and Orion will be back in its original position at midnight. It is interesting to realize that if we were standing on the North Pole we would always see the same part of the sky, since the Earth's rotation will only cause us to turn around our own axis. Then the sky rotates around the point directly above us, the point in the sky we usually refer to as the zenith, and stars only move parallel to the horizon, they do not rise or set. The celestial North Pole is at the zenith. At the South Pole we would see a similar situation, but with only the southern part of the sky visible. Since the axis of our planet does not change its position relative to the stars when it moves around the Sun, the sky visible from the poles of the Earth remains the same all year long or, better, during about six months of darkness. The rest of the year the poles have continuous daylight.

On the other hand, on the Equator we shall see a quite different situation. The celestial Equator, or the line where the plane of the Earth's Equator cuts the celestial sphere, runs from the east, passing directly overhead, to the west. All the stars and other objects in the sky rise and set, and during the year it is possible to view the entire celestial sphere. Finally, in the intermediate areas the situation is more complicated, as you can see in the illustration (Figure 2). There is an area of the sky that is called *circumpolar*. Stars in this part of the sky are so close to the pole that they never rise or set, but remain above the horizon. In the opposite part of the celestial sphere, there is an equally-sized part that never becomes visible.

In the sky maps you can see a band of a slightly lighter blue, representing the brightest parts of the Milky Way. All the stars, nebulae and clusters that

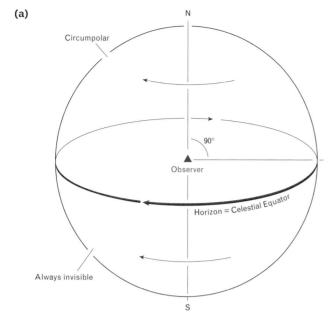

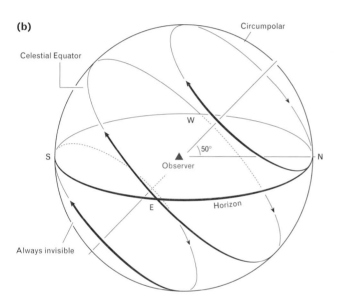

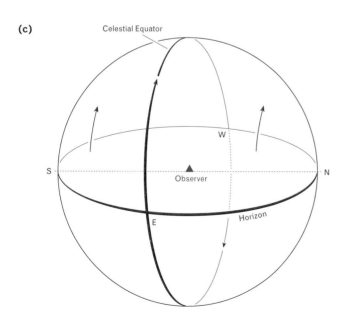

we see in the sky (except the galaxies – see next chapter, where nebulae and clusters are also explained in more detail) belong to a huge formation called the Milky Way Galaxy. Since our Milky Way Galaxy is a large, flat disk-like formation, we see most stars when we look along the plane of the Milky Way. All the light of the millions of stars that we do not see with the naked eye forms a cloud-like band that is easily visible when you are outside big cities and have bright skies.

The twenty-four monthly sky maps are constructed on a stereographic projection. Although this projection has the disadvantage that the scale increases from the center outward, its main advantage is that the shapes of the constellations and star groups are not distorted. When looking at the sky, there is a strange phenomenon, called the Moon illusion, which makes objects appear larger when close to the horizon than when they are high in the sky. This is, however, only a trick of the mind. We believe objects near the horizon to be at a greater distance than objects directly above our heads. The stereographic projection is in a way in harmony with this: constellations near the horizon appear larger!

Choosing the right map

There are twenty-four maps, two for each month, placed side-by-side on facing pages. The map on the left-hand page is for observers in the northern hemisphere and that on the right-hand page for southern observers. Each map has four different horizons, labeled to show you for what latitude on Earth the horizon is exact. Depending on the latitude on Earth at which you are living, you can select your horizon. A few degrees more or less will not make too much difference, when you are looking at the sky. The

Figure 2 (a) The way the sky moves for an observer at the North Pole. Half of the sky is always visible (circumpolar), and the other half is always invisible. Stars do not rise or set.
(b) The way the sky moves for an observer at a latitude of 50° north. A smaller part of the northern sky is always visible (circumpolar), and a similar area of the southern sky is always invisible. Most stars rise and set.
(c) The way the sky moves for an observer at the Equator. There are no circumpolar stars, and no stars that are always invisible. All stars rise and set.

Table A *Selecting the monthly sky maps*

		5 PM	6 PM	7 PM	8 PM	9 PM	10 PM	11 PM	Midnight	1 AM	2 AM	3 AM	4 AM	5 AM	6 AM	7 AM
January	1	Oct		Nov		Dec		Jan		Feb		Mar		Apr		May
January	15		Nov		Dec		Jan		Feb		Mar		Apr		May	
February	1	Nov		Dec		Jan		Feb		Mar		Apr		May		Jun
February	15		Dec		Jan		Feb		Mar		Apr		May		Jun	
March	1	Dec		Jan		Feb		Mar		Apr		May		Jun		Jul
March	15		Jan		Feb		Mar		Apr		May		Jun		Jul	
April	1	Jan		Feb		Mar		Apr		May		Jun		Jul		Aug
April	15		Feb		Mar		Apr		May		Jun		Jul		Aug	
May	1	Feb		Mar		Apr		May		Jun		Jul		Aug		Sep
May	15		Mar		Apr		May		Jun		Jul		Aug		Sep	
June	1	Mar		Apr		May		Jun		Jul		Aug		Sep		Oct
June	15		Apr		May		Jun		Jul		Aug		Sep		Oct	
July	1	Apr		May		Jun		Jul		Aug		Sep		Oct		Nov
July	15		May		Jun		Jul		Aug		Sep		Oct		Nov	
August	1	May		Jun		Jul		Aug		Sep		Oct		Nov		Dec
August	15		Jun		Jul		Aug		Sep		Oct		Nov		Dec	
September	1	Jun		Jul		Aug		Sep		Oct		Nov		Dec		Jan
September	15		Jul		Aug		Sep		Oct		Nov		Dec		Jan	
October	1	Jul		Aug		Sep		Oct		Nov		Dec		Jan		Feb
October	15		Aug		Sep		Oct		Nov		Dec		Jan		Feb	
November	1	Aug		Sep		Oct		Nov		Dec		Jan		Feb		Mar
November	15		Sep		Oct		Nov		Dec		Jan		Feb		Mar	
December	1	Sep		Oct		Nov		Dec		Jan		Feb		Mar		Apr
December	15		Oct		Nov		Dec		Jan		Feb		Mar		Apr	

maps are for 11 PM on the first day of the given month, 10 PM on the 15th and 9 PM on the first day of the following month. These times are given at the bottom (left) of each map. But the maps can be used over a range of dates and times, as is shown in Table A above. Let's say you want to look at the sky in the middle of January, not in the evening, but at 6 AM. Simply look for January 15 in the left-hand column, and then look under 6 AM. You will find that

you have to use the maps for the month of May. When local Summer Time (Daylight Saving Time; DST) is used, one hour should be added to the times given in the table (or one hour subtracted from the time on your watch).

The key for the star magnitudes is shown at the bottom right of each map. The word magnitude is explained in the introduction to the star charts, on page 33.

Northern latitudes

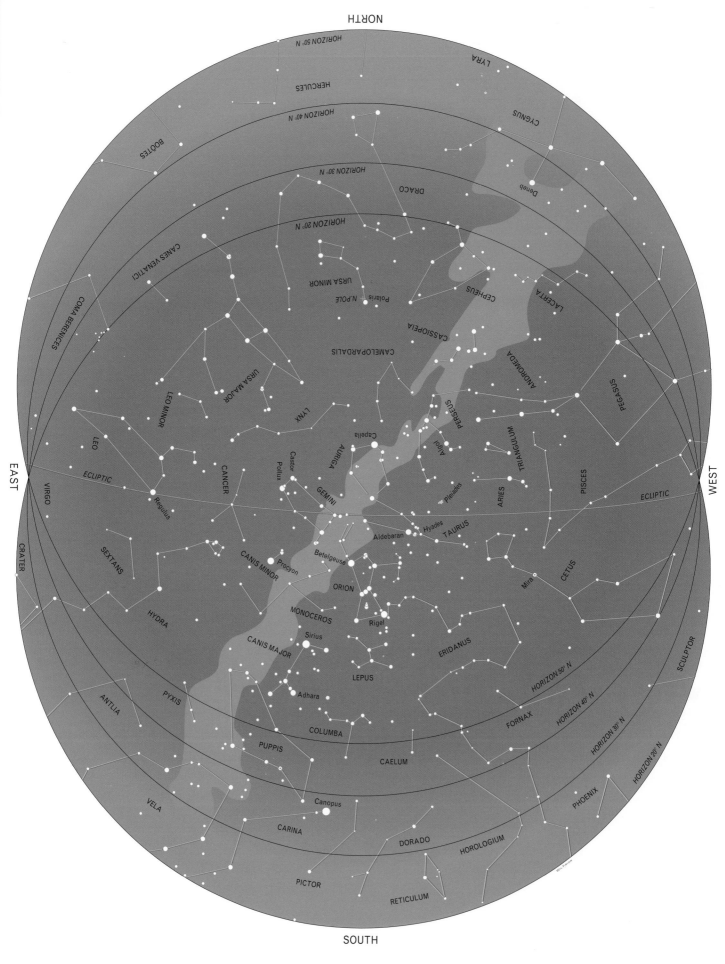

NORTH

HORIZON 50° N

HORIZON 40° N

HORIZON 30° N

HORIZON 20° N

LYRA

HERCULES

CYGNUS

BOÖTES

Deneb

DRACO

CANES VENATICI

URSA MINOR

CEPHEUS

LACERTA

COMA BERENICES

N.POLE
Polaris

CASSIOPEIA

ANDROMEDA

PEGASUS

CAMELOPARDALIS

LEO MINOR

URSA MAJOR

LYNX

PERSEUS

TRIANGULUM

LEO

AURIGA
Capella

Algol

PISCES

EAST

CANCER

Castor
Pollux

GEMINI

Pleiades

ARIES

WEST

ECLIPTIC

VIRGO

Regulus

Aldebaran

Hyades

TAURUS

ECLIPTIC

CRATER

SEXTANS

CANIS MINOR
Procyon

Betelgeuse

CETUS

Mira

ORION

MONOCEROS

Rigel

HYDRA

CANIS MAJOR

ERIDANUS

Sirius

LEPUS

HORIZON 50° N

SCULPTOR

PYXIS

Adhara

HORIZON 40° N

ANTLIA

COLUMBA

CAELUM

FORNAX

HORIZON 30° N

PUPPIS

HORIZON 20° N

PHOENIX

VELA

Canopus

CARINA

DORADO

HOROLOGIUM

PICTOR

RETICULUM

SOUTH

Date	Time	DST
January 1	11 pm	Midnight
January 15	10 pm	11 pm
February 1	9 pm	10 pm

Magnitudes:

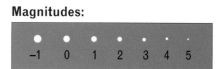

-1   0   1   2   3   4   5

Southern latitudes

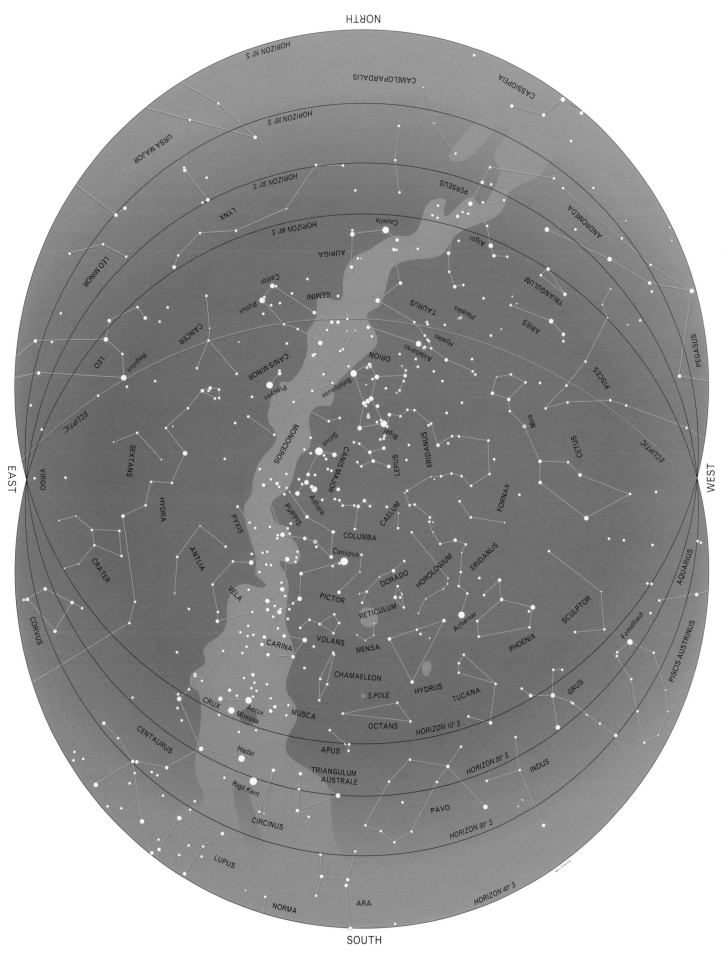

Date	Time	DST
January 1	11 pm	Midnight
January 15	10 pm	11 pm
February 1	9 pm	10 pm

Magnitudes:

−1   0   1   2   3   4   5

Northern latitudes

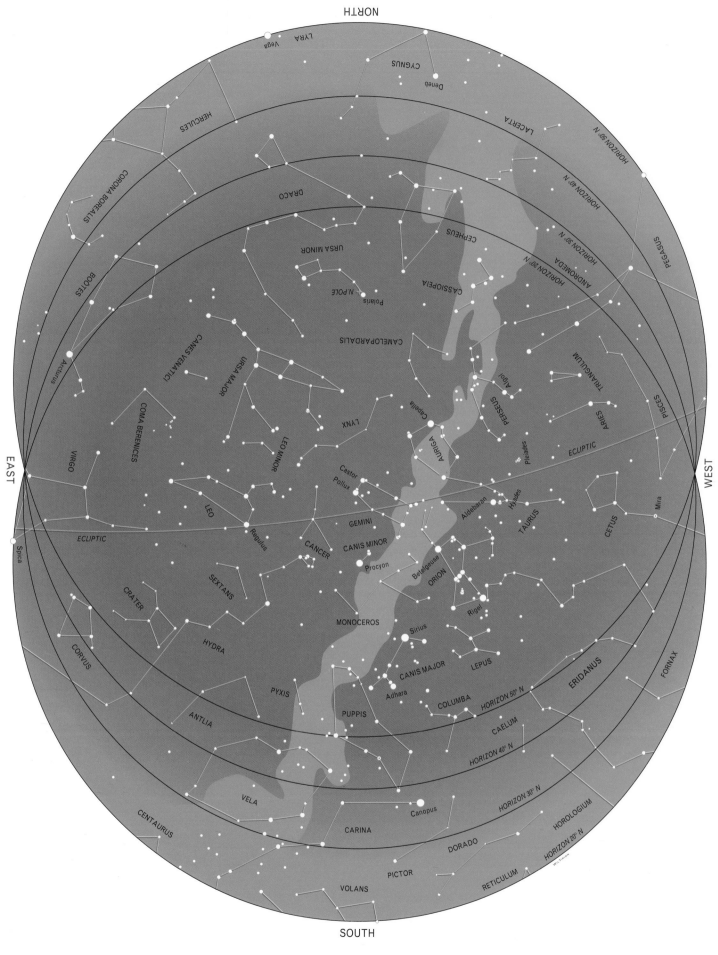

NORTH

LYRA
Vega
CYGNUS
Deneb
LACERTA
HERCULES
HORIZON 50° N
CORONA BOREALIS
DRACO
CEPHEUS
HORIZON 40° N
PEGASUS
BOÖTES
URSA MINOR
CASSIOPEIA
HORIZON 30° N
N.POLE
Polaris
ANDROMEDA
HORIZON 20° N
Arcturus
CANES VENATICI
CAMELOPARDALIS
TRIANGULUM
PISCES
COMA BERENICES
URSA MAJOR
Algol
ARIES
VIRGO
LYNX
PERSEUS
Pleiades
ECLIPTIC
Capella
AURIGA
Hyades
LEO MINOR
Aldebaran
TAURUS
Castor
Pollux
LEO
Regulus
GEMINI
CETUS
Mira
CANCER
CANIS MINOR
Betelgeuse
ECLIPTIC
SEXTANS
Procyon
ORION
Spica
CRATER
MONOCEROS
Rigel
HYDRA
PYXIS
CORVUS
ANTLIA
Sirius
CANIS MAJOR
LEPUS
ERIDANUS
FORNAX
Adhara
PUPPIS
COLUMBA
HORIZON 50° N
VELA
CAELUM
HORIZON 40° N
CENTAURUS
Canopus
CARINA
HORIZON 30° N
HOROLOGIUM
DORADO
HORIZON 20° N
PICTOR
RETICULUM
VOLANS

EAST

WEST

SOUTH

Date	Time	DST
February 1	11 pm	Midnight
February 15	10 pm	11 pm
March 1	9 pm	10 pm

Magnitudes:

−1 0 1 2 3 4 5

Southern latitudes

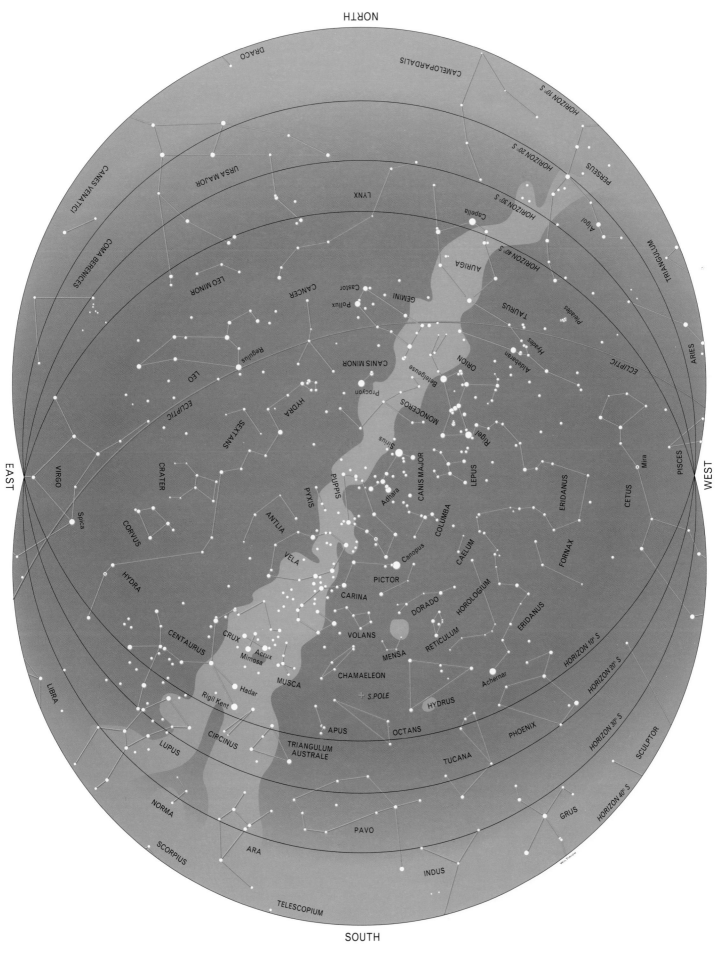

Date	Time	DST
February 1	11 pm	Midnight
February 15	10 pm	11 pm
March 1	9 pm	10 pm

Magnitudes:

−1	0	1	2	3	4	5

Northern latitudes

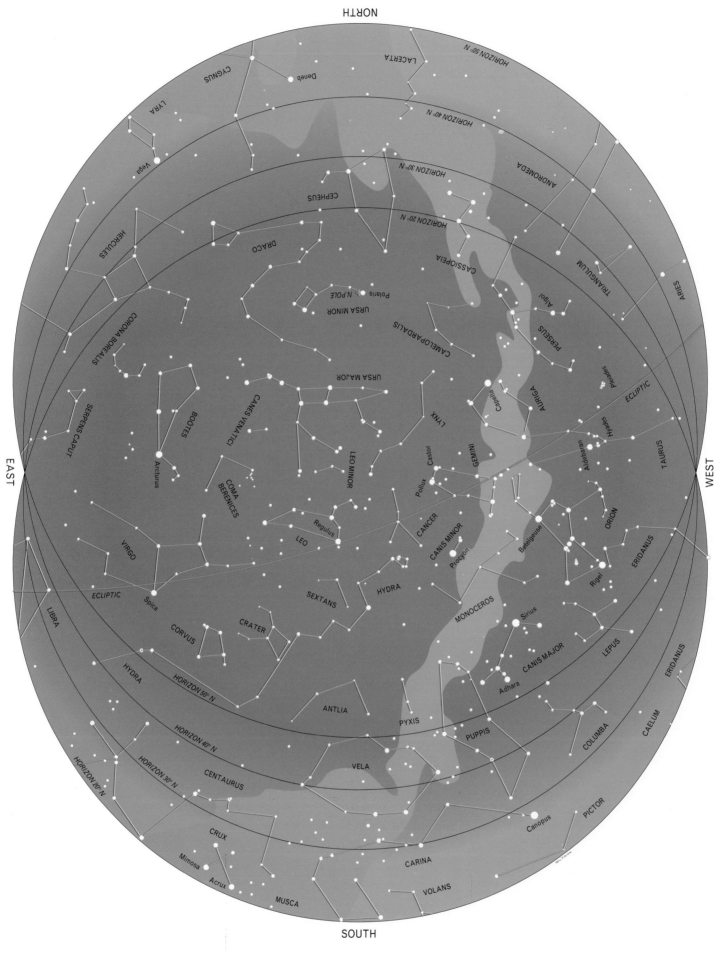

Date	Time	DST
March 1	11 pm	Midnight
March 15	10 pm	11 pm
April 1	9 pm	10 pm

Magnitudes:

−1 0 1 2 3 4 5

Southern latitudes

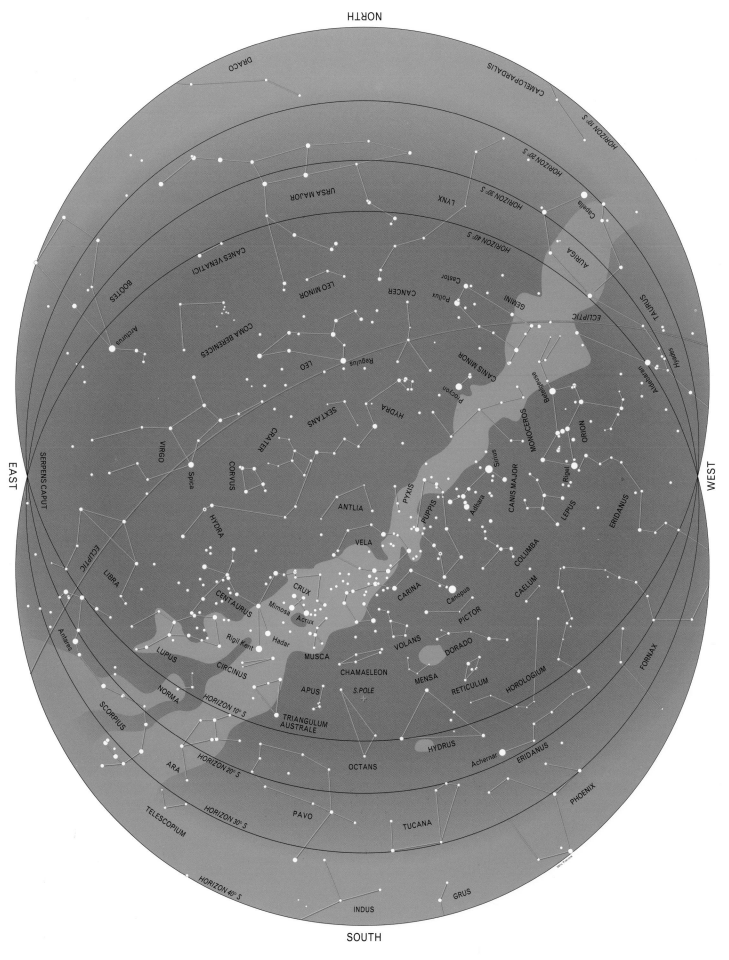

NORTH

DRACO

CAMELOPARDALIS

HORIZON 10° S

URSA MAJOR

LYNX

HORIZON 20° S

Capella

CANES VENATICI

HORIZON 30° S

LEO MINOR

CANCER

Castor

HORIZON 40° S

Pollux

GEMINI

AURIGA

BOOTES

COMA BERENICES

Castor

TAURUS

Arcturus

LEO

Regulus

CANIS MINOR

ECLIPTIC

Aldebaran

Hyades

SERPENS CAPUT

SEXTANS

HYDRA

Procyon

Betelgeuse

ORION

EAST

VIRGO

CRATER

Spica

ANTLIA

PYXIS

MONOCEROS

Rigel

WEST

ECLIPTIC

CORVUS

PUPPIS

Sirius

CANIS MAJOR

LEPUS

ERIDANUS

LIBRA

HYDRA

VELA

Adhara

COLUMBA

CRUX

CARINA

CAELUM

CENTAURUS

Mimosa

Acrux

Canopus

Antares

Rigil Kent

Hadar

MUSCA

VOLANS

PICTOR

DORADO

FORNAX

LUPUS

CIRCINUS

CHAMAELEON

MENSA

RETICULUM

HOROLOGIUM

NORMA

APUS

S. POLE

SCORPIUS

HORIZON 10° S

TRIANGULUM
AUSTRALE

HYDRUS

ARA

HORIZON 20° S

OCTANS

Achernar

ERIDANUS

TELESCOPIUM

HORIZON 30° S

PAVO

PHOENIX

TUCANA

HORIZON 40° S

GRUS

INDUS

Will Tirion

SOUTH

Date	Time	DST
March 1	11 pm	Midnight
March 15	10 pm	11 pm
April 1	9 pm	10 pm

Magnitudes:

−1	0	1	2	3	4	5

Northern latitudes

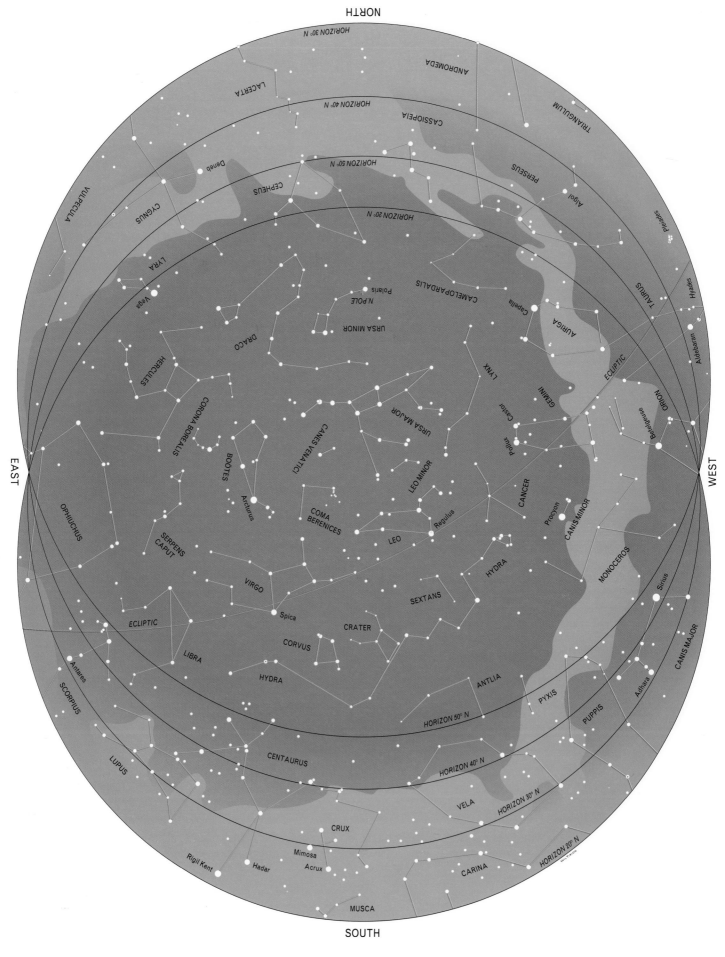

Date	Time	DST
April 1	11 pm	Midnight
April 15	10 pm	11 pm
May 1	9 pm	10 pm

Magnitudes:

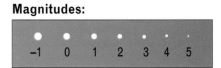

−1   0   1   2   3   4   5

Southern latitudes

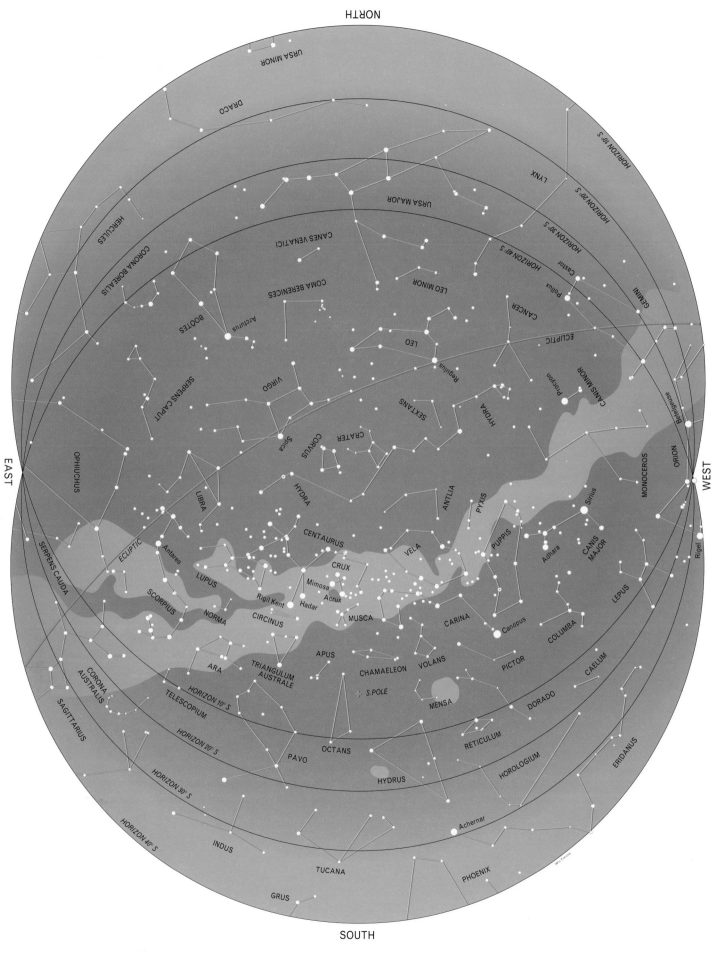

NORTH

URSA MINOR

DRACO

HERCULES

URSA MAJOR

CORONA BOREALIS

CANES VENATICI

LYNX

HORIZON 10° S

HORIZON 20° S

HORIZON 30° S

COMA BERENICES

LEO MINOR

HORIZON 40° S

Castor

GEMINI

BOÖTES

Arcturus

CANCER

Pollux

SERPENS CAPUT

VIRGO

LEO

ECLIPTIC

CANIS MINOR

Regulus

Procyon

Betelgeuse

EAST

OPHIUCHUS

Spica

CORVUS

CRATER

SEXTANS

HYDRA

ORION

WEST

HYDRA

MONOCEROS

Rigel

LIBRA

ANTLIA

PYXIS

Sirius

CENTAURUS

VELA

PUPPIS

Adhara

CANIS MAJOR

SERPENS CAUDA

ECLIPTIC

Antares

CRUX

CARINA

LEPUS

Mimosa

Acrux

SCORPIUS

LUPUS

Rigil Kent

Hadar

MUSCA

Canopus

COLUMBA

NORMA

CIRCINUS

CARINA

PICTOR

CAELUM

ARA

APUS

VOLANS

TRIANGULUM

AUSTRALE

CHAMAELEON

CORONA

AUSTRALIS

S.POLE

MENSA

DORADO

ERIDANUS

TELESCOPIUM

HORIZON 10° S

OCTANS

RETICULUM

SAGITTARIUS

HORIZON 20° S

PAVO

HOROLOGIUM

HORIZON 30° S

HYDRUS

Achernar

HORIZON 40° S

INDUS

PHOENIX

TUCANA

GRUS

SOUTH

Date	Time	DST
April 1	11 pm	Midnight
April 15	10 pm	11 pm
May 1	9 pm	10 pm

Magnitudes:

−1 0 1 2 3 4 5

April

Northern latitudes

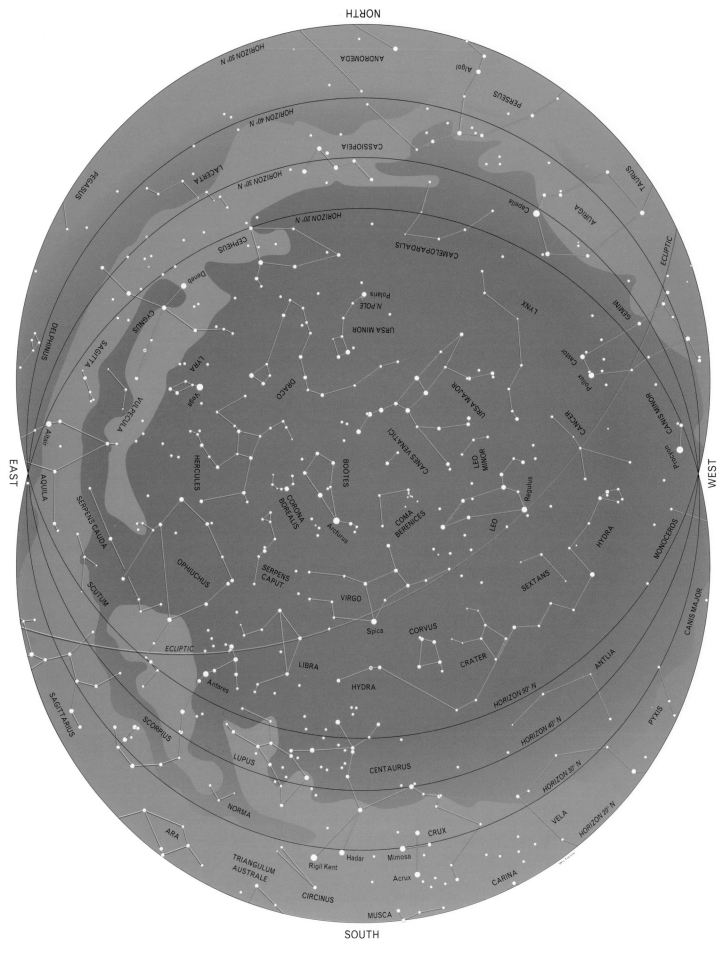

NORTH

HORIZON 50° N
HORIZON 40° N
HORIZON 30° N
HORIZON 20° N

ANDROMEDA
Algol
PERSEUS
CASSIOPEIA
TAURUS
LACERTA
PEGASUS
CEPHEUS
Capella
AURIGA
CAMELOPARDALIS
GEMINI
Deneb
CYGNUS
N.POLE
Polaris
URSA MINOR
LYNX
ECLIPTIC
DELPHINUS
SAGITTA
VULPECULA
LYRA
DRACO
Pollux
Castor
CANIS MINOR
Vega
URSA MAJOR
CANCER
Procyon
Altair
HERCULES
BOOTES
CANES VENATICI
LEO MINOR
AQUILA
CORONA
BOREALIS
Arcturus
COMA
BERENICES
LEO
Regulus
HYDRA
SERPENS CAUDA
OPHIUCHUS
SERPENS
CAPUT
VIRGO
SEXTANS
MONOCEROS
SCUTUM
Spica
CORVUS
CRATER
CANIS MAJOR
ECLIPTIC
LIBRA
HYDRA
ANTLIA
SAGITTARIUS
Antares
HORIZON 50° N
PYXIS
SCORPIUS
LUPUS
CENTAURUS
HORIZON 40° N
NORMA
HORIZON 30° N
VELA
ARA
HORIZON 20° N
CRUX
TRIANGULUM
AUSTRALE
Hadar
Mimosa
CARINA
Rigil Kent
CIRCINUS
Acrux
MUSCA

SOUTH

EAST

WEST

Date	Time	DST
May 1	11 pm	Midnight
May 15	10 pm	11 pm
June 1	9 pm	10 pm

Magnitudes:

−1 0 1 2 3 4 5

Southern latitudes

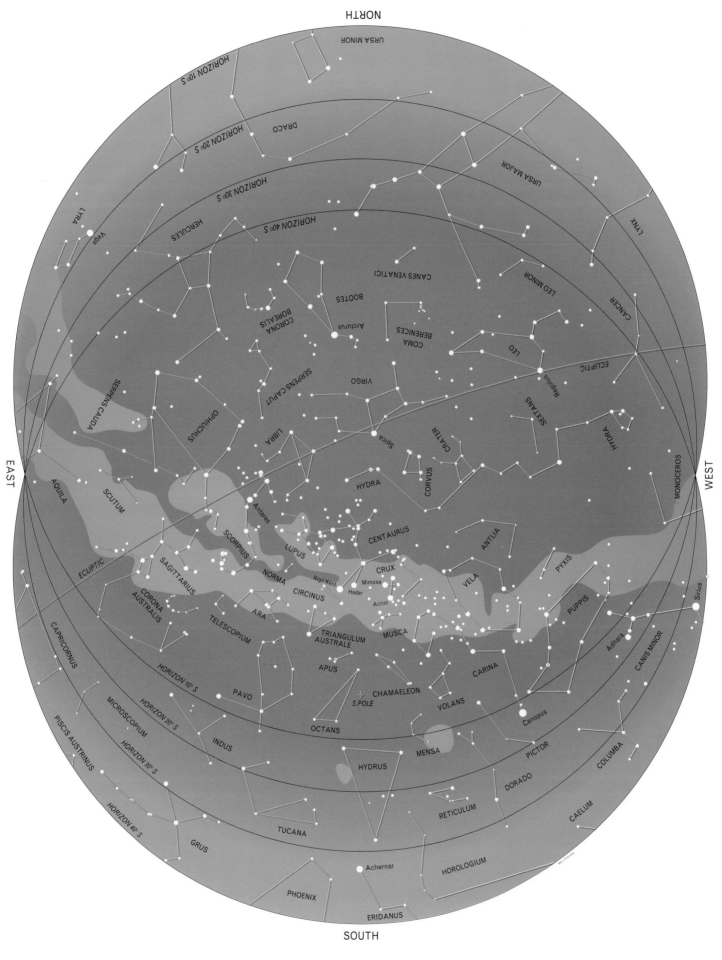

Date	Time	DST
May 1	11 pm	Midnight
May 15	10 pm	11 pm
June 1	9 pm	10 pm

Magnitudes:

−1   0   1   2   3   4   5

Northern latitudes

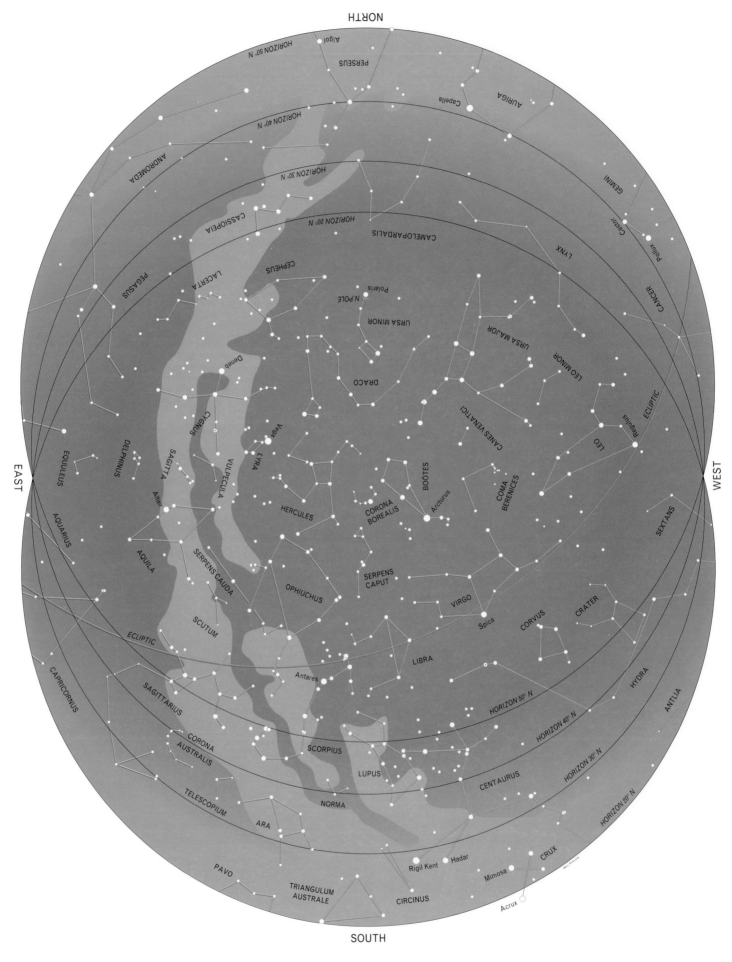

Date	Time	DST
June 1	11 pm	Midnight
June 15	10 pm	11 pm
July 1	9 pm	10 pm

Magnitudes:

−1   0   1   2   3   4   5

Southern latitudes

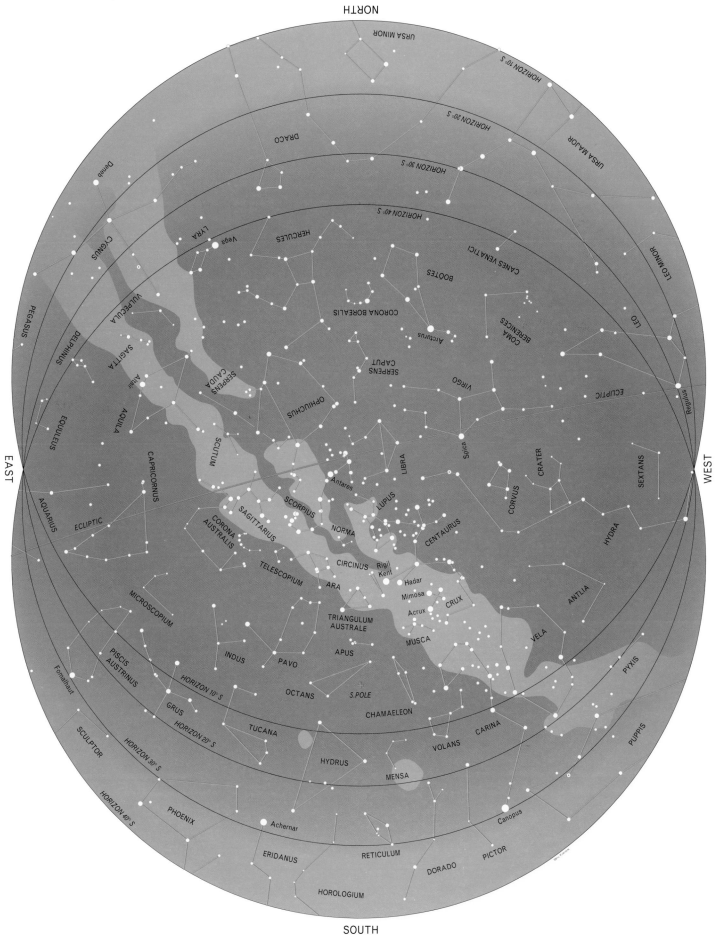

NORTH

URSA MINOR

HORIZON 10° S

URSA MAJOR

HORIZON 20° S

DRACO

HORIZON 30° S

LEO MINOR

HORIZON 40° S

Denab

LYRA

HERCULES

Vega

CYGNUS

CANES VENATICI

BOOTES

CORONA BOREALIS

Arcturus

LEO

PEGASUS

VULPECULA

COMA BERENICES

DELPHINUS

SAGITTA

SERPENS CAPUT

ECLIPTIC

EQUULEUS

AQUILA

Altair

SCUTUM

SERPENS CAUDA

VIRGO

Regulus

Spica

WEST

EAST

AQUARIUS

OPHIUCHUS

LIBRA

CRATER

SEXTANS

CAPRICORNUS

ECLIPTIC

Antares

SCORPIUS

LUPUS

CENTAURUS

CORVUS

HYDRA

CORONA AUSTRALIS

SAGITTARIUS

NORMA

ANTLIA

MICROSCOPIUM

TELESCOPIUM

CIRCINUS

Rigil Kent

Hadar

Mimosa

CRUX

VELA

ARA

Acrux

MUSCA

PYXIS

INDUS

PAVO

TRIANGULUM AUSTRALE

APUS

PISCIS AUSTRINUS

HORIZON 10° S

OCTANS

S. POLE

Fomalhaut

GRUS

HORIZON 20° S

TUCANA

CHAMAELEON

CARINA

PUPPIS

SCULPTOR

HORIZON 30° S

VOLANS

HYDRUS

MENSA

HORIZON 40° S

PHOENIX

Achernar

Canopus

RETICULUM

PICTOR

ERIDANUS

DORADO

HOROLOGIUM

SOUTH

Date	Time	DST
June 1	11 pm	Midnight
June 15	10 pm	11 pm
July 1	9 pm	10 pm

Magnitudes:

−1 0 1 2 3 4 5

June

Northern latitudes

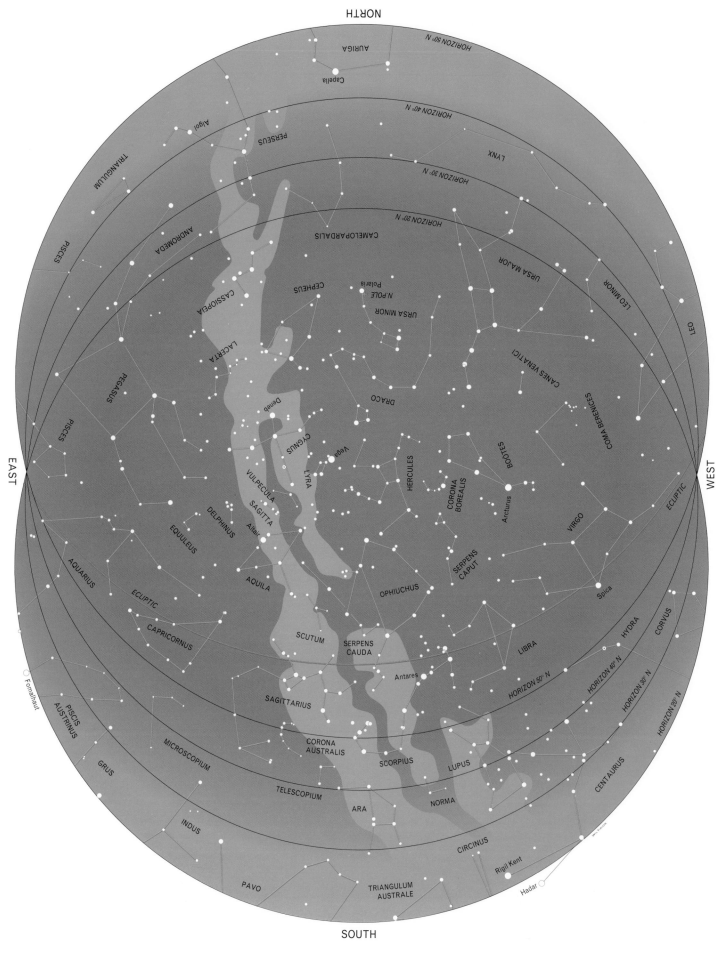

NORTH

HORIZON 50° N
HORIZON 40° N
HORIZON 30° N
HORIZON 20° N

AURIGA
Capella
Algol
PERSEUS
TRIANGULUM
PISCES
PISCES
ANDROMEDA
PEGASUS
CASSIOPEIA
LACERTA
Deneb
CYGNUS
VULPECULA
SAGITTA
Altair
DELPHINUS
EQUULEUS
AQUILA
AQUARIUS
ECLIPTIC
CAPRICORNUS
SCUTUM
SERPENS CAUDA

LYNX
CAMELOPARDALIS
CEPHEUS
N.POLE
Polaris
URSA MINOR
URSA MAJOR
LEO MINOR
LEO
CANES VENATICI
DRACO
COMA BERENICES
BOOTES
Vega
LYRA
HERCULES
CORONA BOREALIS
Arcturus
SERPENS CAPUT
OPHIUCHUS
VIRGO
Spica
LIBRA
HYDRA
CORVUS
Antares
HORIZON 40° N
HORIZON 30° N
HORIZON 20° N

EAST
WEST
ECLIPTIC

Fomalhaut
PISCIS AUSTRINUS
GRUS
MICROSCOPIUM
INDUS
SAGITTARIUS
CORONA AUSTRALIS
SCORPIUS
LUPUS
TELESCOPIUM
ARA
NORMA
CENTAURUS
CIRCINUS
Rigil Kent
PAVO
TRIANGULUM AUSTRALE
Hadar

HORIZON 50° N

SOUTH

Date	Time	DST
July 1	11 pm	Midnight
July 15	10 pm	11 pm
August 1	9 pm	10 pm

Magnitudes:

−1 0 1 2 3 4 5

Southern latitudes

Date	Time	DST
July 1	11 pm	Midnight
July 15	10 pm	11 pm
August 1	9 pm	10 pm

Magnitudes:

−1 0 1 2 3 4 5

Northern latitudes

Date	Time	DST
August 1	11 pm	Midnight
August 15	10 pm	11 pm
September 1	9 pm	10 pm

Magnitudes:

| −1 | 0 | 1 | 2 | 3 | 4 | 5 |

Southern latitudes

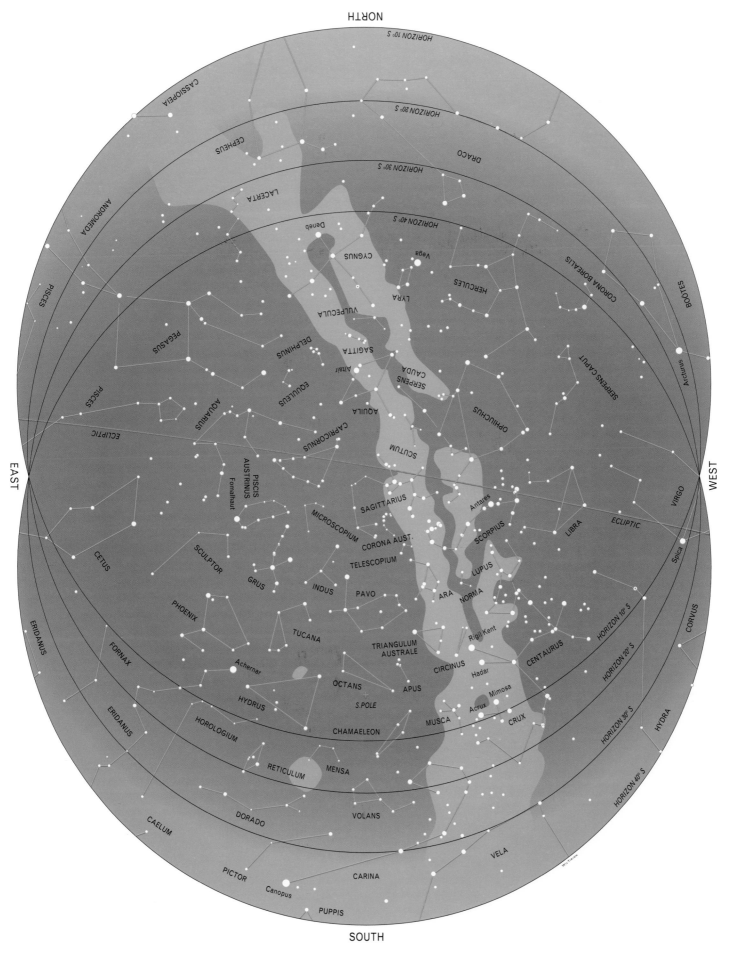

Date	Time	DST
August 1	11 pm	Midnight
August 15	10 pm	11 pm
September 1	9 pm	10 pm

Magnitudes:

| −1 | 0 | 1 | 2 | 3 | 4 | 5 |

Northern latitudes

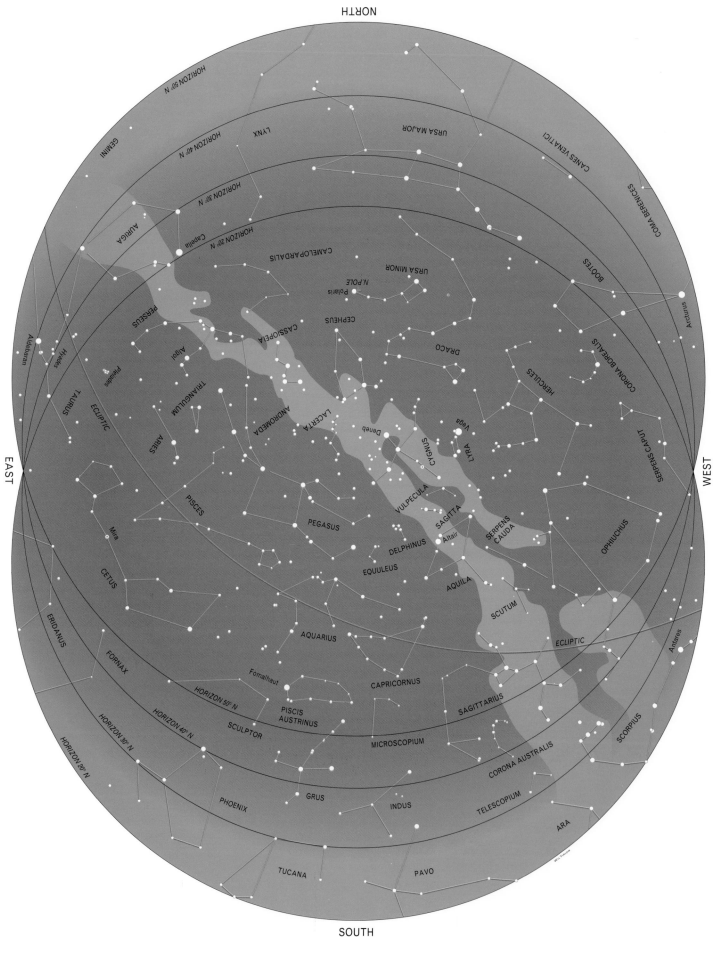

NORTH

HORIZON 50° N
HORIZON 40° N
HORIZON 30° N
HORIZON 20° N

GEMINI
LYNX
URSA MAJOR
CANES VENATICI
COMA BERENICES

AURIGA
Capella
CAMELOPARDALIS
URSA MINOR
Polaris
N.POLE
DRACO
BOÖTES
Arcturus

PERSEUS
Algol
CASSIOPEIA
CEPHEUS
HERCULES
SERPENS CAPUT
CORONA BOREALIS

Aldebaran
Hyades
Pleiades
TRIANGULUM
ANDROMEDA
LACERTA
Deneb
CYGNUS
Vega
LYRA

TAURUS
ECLIPTIC
ARIES
PISCES
PEGASUS
VULPECULA
SAGITTA
SERPENS CAUDA
OPHIUCHUS

EAST
WEST

Mira
Altair
DELPHINUS
AQUILA
SCUTUM

CETUS
EQUULEUS

ERIDANUS
AQUARIUS
CAPRICORNUS
ECLIPTIC
Antares

FORNAX
Fomalhaut
SAGITTARIUS
SCORPIUS

HORIZON 50° N
MICROSCOPIUM
CORONA AUSTRALIS

HORIZON 40° N
SCULPTOR
PISCIS AUSTRINUS

HORIZON 30° N
GRUS
INDUS
TELESCOPIUM
ARA

HORIZON 20° N
PHOENIX

TUCANA
PAVO

SOUTH

Date	Time	DST
September 1	11 pm	Midnight
September 15	10 pm	11 pm
October 1	9 pm	10 pm

Magnitudes:

| -1 | 0 | 1 | 2 | 3 | 4 | 5 |

Southern latitudes

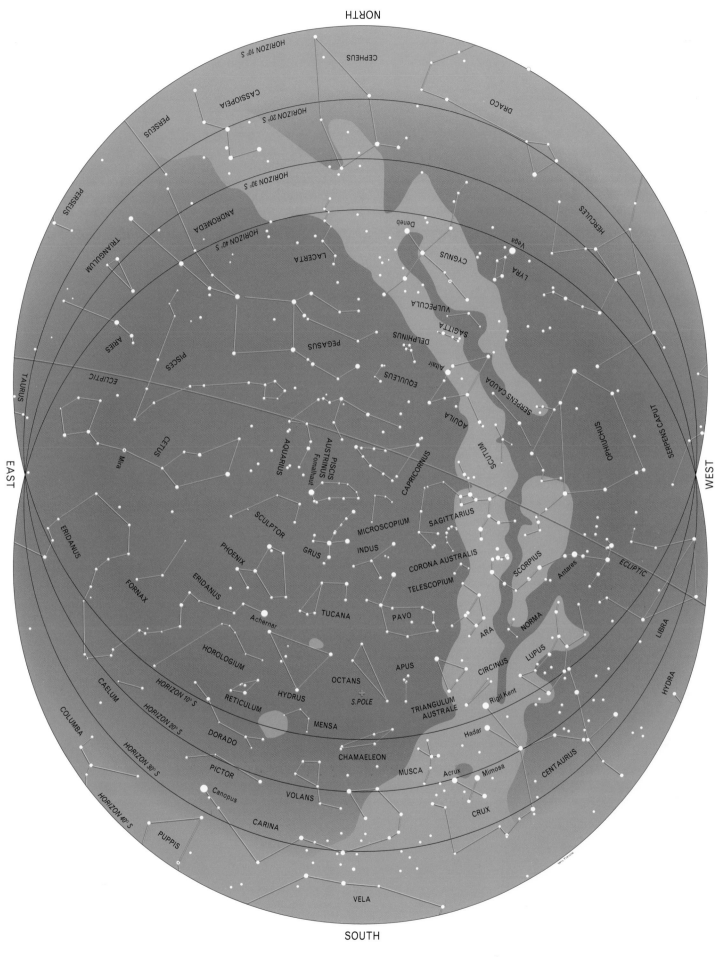

Date	Time	DST
September 1	11 pm	Midnight
September 15	10 pm	11 pm
October 1	9 pm	10 pm

Magnitudes:

−1 0 1 2 3 4 5

Northern latitudes

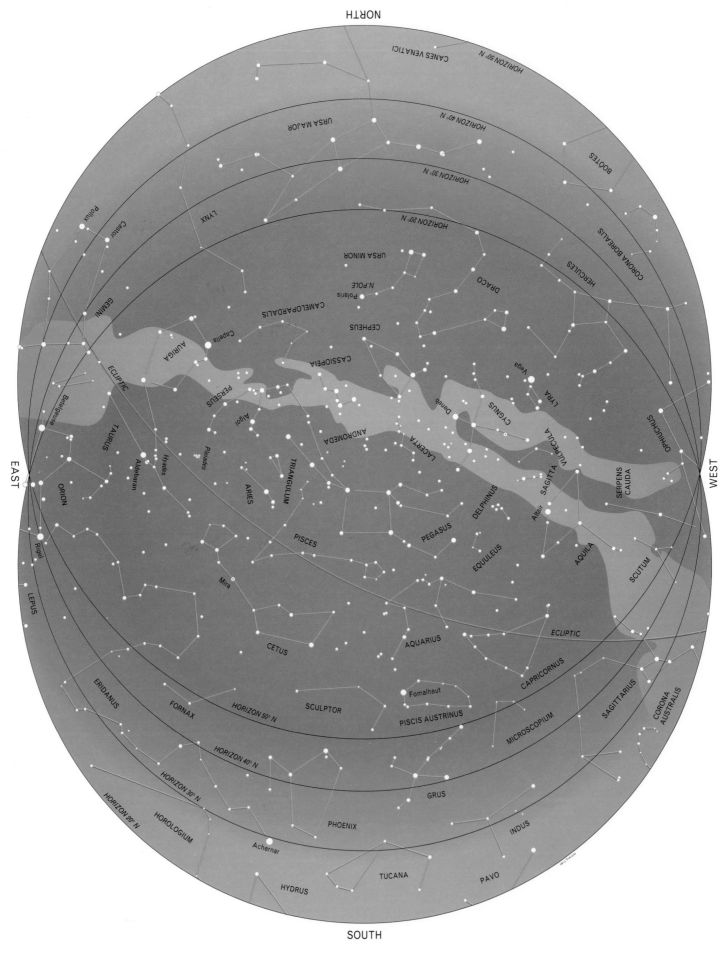

Date	Time	DST
October 1	11 pm	Midnight
October 15	10 pm	11 pm
November 1	9 pm	10 pm

Magnitudes:

−1	0	1	2	3	4	5

Southern latitudes

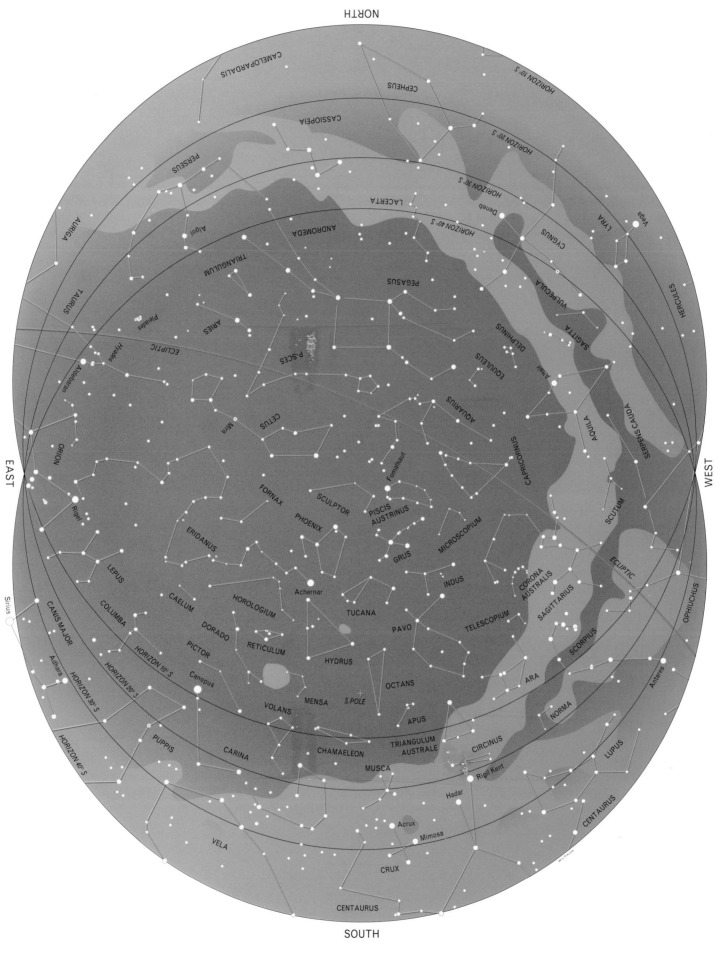

Date	Time	DST
October 1	11 pm	Midnight
October 15	10 pm	11 pm
November 1	9 pm	10 pm

Magnitudes:

-1 0 1 2 3 4 5

Northern latitudes

NORTH

HORIZON 50° N

CANES VENATICI

BOÖTES

HORIZON 40° N

LEO MINOR

URSA MAJOR

HORIZON 30° N

HERCULES

LEO

HORIZON 20° N

DRACO

ECLIPTIC

LYNX

URSA MINOR

N.POLE

Polaris

CEPHEUS

Vega

LYRA

CYGNUS

CANCER

CAMELOPARDALIS

CASSIOPEIA

Deneb

VULPECULA

SAGITTA

Castor

Pollux

AURIGA

Capella

PERSEUS

Algol

LACERTA

Altair

AQUILA

GEMINI

Procyon

ANDROMEDA

DELPHINUS

WEST

EAST

CANIS MINOR

Pleiades

TRIANGULUM

PEGASUS

EQUULEUS

Betelgeuse

Hyades

ARIES

MONOCEROS

Aldebaran

PISCES

ORION

TAURUS

ECLIPTIC

Rigel

Mira

CETUS

CAPRICORNUS

Sirius

AQUARIUS

ERIDANUS

CANIS MAJOR

LEPUS

Fomalhaut

FORNAX

SCULPTOR

PISCIS AUSTRINUS

MICROSCOPIUM

COLUMBA

HORIZON 50° N

CAELUM

PHOENIX

HORIZON 40° N

GRUS

PICTOR

HOROLOGIUM

HORIZON 30° N

INDUS

DORADO

Achernar

TUCANA

RETICULUM

HORIZON 20° N

HYDRUS

SOUTH

Date	Time	DST
November 1	11 pm	Midnight
November 15	10 pm	11 pm
December 1	9 pm	10 pm

Magnitudes:

−1	0	1	2	3	4	5

Southern latitudes

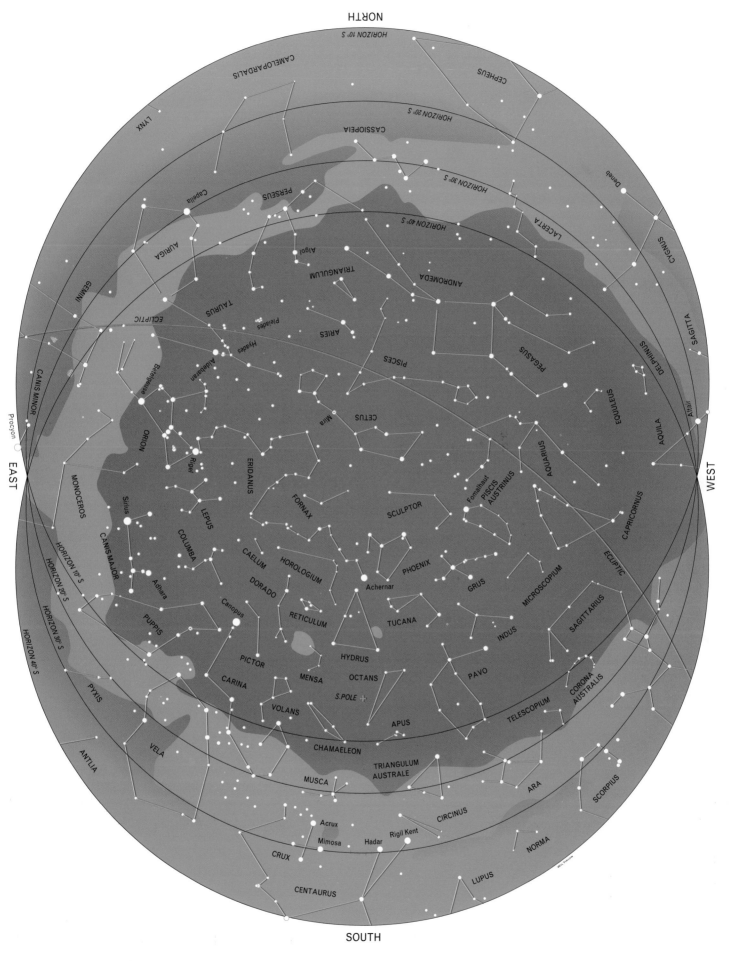

Date	Time	DST
November 1	11 pm	Midnight
November 15	10 pm	11 pm
December 1	9 pm	10 pm

Magnitudes:

−1 0 1 2 3 4 5

Northern latitudes

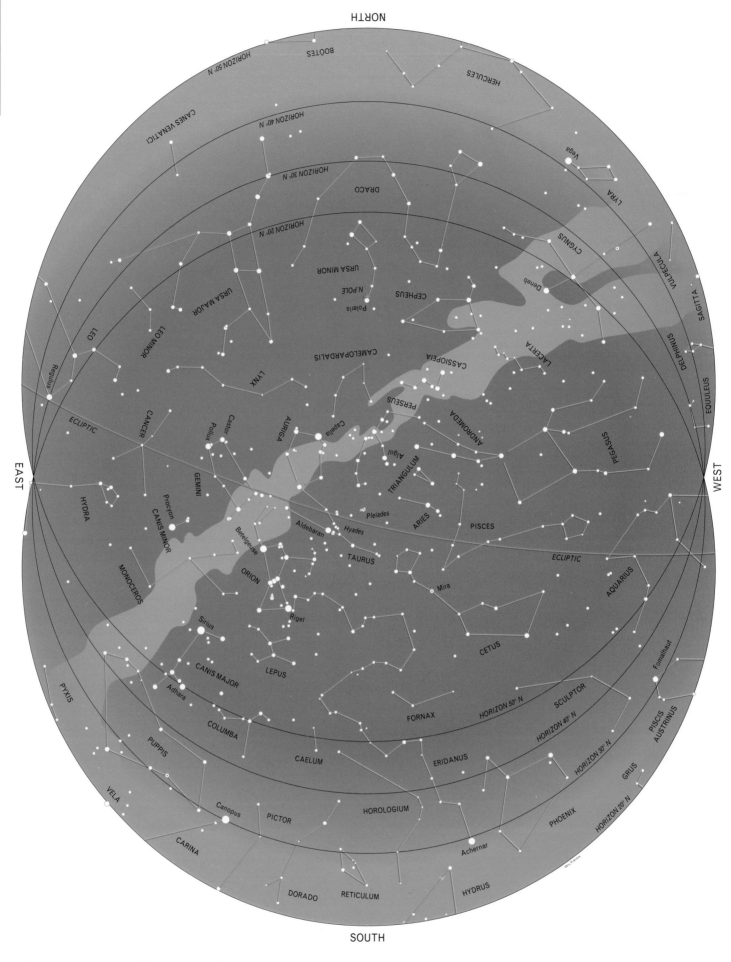

NORTH

HORIZON 50° N
BOOTES
HERCULES
HORIZON 40° N
CANES VENATICI
Vega
DRACO
HORIZON 30° N
LYRA
CYGNUS
HORIZON 20° N
URSA MINOR
N.POLE
CEPHEUS
Deneb
VULPECULA
Polaris
CASSIOPEIA
LACERTA
SAGITTA
CAMELOPARDALIS
URSA MAJOR
PERSEUS
DELPHINUS
LEO MINOR
LEO
LYNX
ANDROMEDA
PEGASUS
EQUULEUS
Capella
Algol
Regulus
AURIGA
TRIANGULUM
ECLIPTIC
CANCER
Castor
Pollux
ARIES
PISCES
Pleiades

EAST   **WEST**

GEMINI
Procyon
Hyades
HYDRA
CANIS MINOR
Aldebaran
Betelgeuse
Mira
TAURUS
AQUARIUS
MONOCEROS
ORION
ECLIPTIC
Rigel
Sirius
LEPUS
CETUS
CANIS MAJOR
PYXIS
Adhara
Fomalhaut
COLUMBA
FORNAX
HORIZON 50° N
SCULPTOR
PISCIS
AUSTRINUS
PUPPIS
CAELUM
HORIZON 40° N
ERIDANUS
GRUS
VELA
PICTOR
HORIZON 30° N
PHOENIX
Canopus
HOROLOGIUM
HORIZON 20° N
CARINA
Achernar
DORADO
RETICULUM
HYDRUS

SOUTH

Date	Time	DST
December 1	11 pm	Midnight
December 15	10 pm	11 pm
January 1	9 pm	10 pm

Magnitudes:

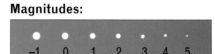

−1 0 1 2 3 4 5

Southern latitudes

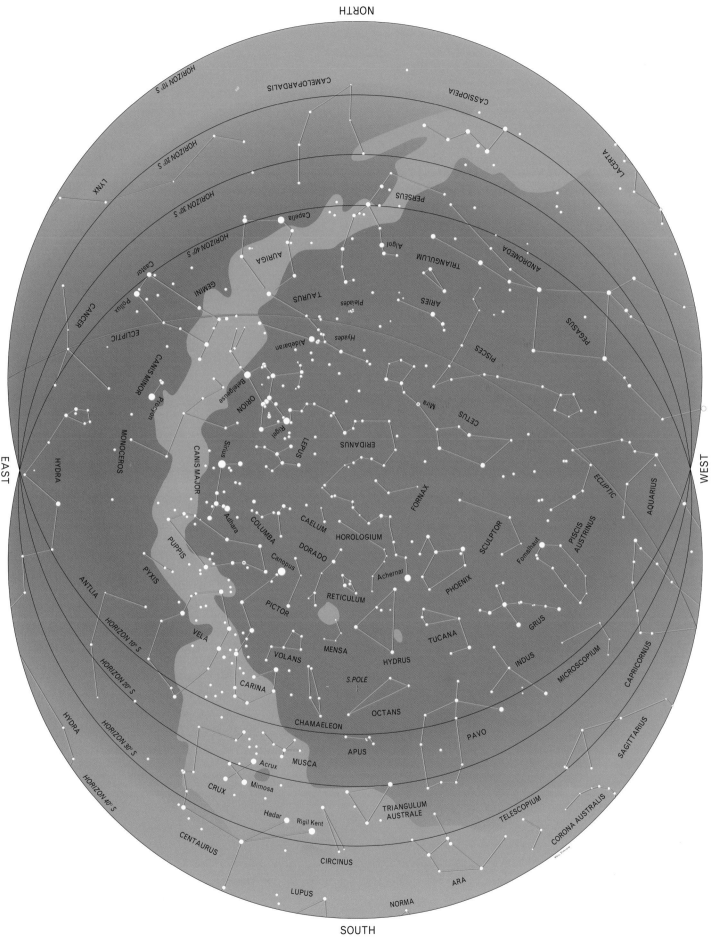

Date	Time	DST
December 1	11 pm	Midnight
December 15	10 pm	11 pm
January 1	9 pm	10 pm

Magnitudes:

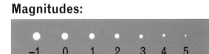

−1 0 1 2 3 4 5

THE STAR CHARTS

Once you are familiar with the sky and can recognize the different constellations, you will want to know more about what is visible in the night sky. So the next step is to use the star charts in this chapter, the 'heart' of the *Cambridge Star Atlas*. These star charts divide the sky into 20 parts. The actual chart areas are shown on the index to the star charts on page 41, preceding the star charts themselves. There is a generous overlap between the charts, so most of the constellations are shown complete on at least one chart. The positions of stars and objects are for the year 2000 or, to be more precise, the epoch is 2000.0; the extra 0 is a decimal and it means January 1. (2000.5 would be July 1.) The positions are plotted against a grid of right ascension (RA) and declination (Dec) comparable with longitude and latitude on the Earth's globe. *Right ascension (RA)* is reckoned in hours, minutes and seconds from 0 h to 24 h, from west to east along the Equator. *Declination (Dec)* represents the angular distance between an object and the celestial Equator, (+) for objects north and (−) for those south of the Equator (Figure 3).

The charts' projections have been carefully chosen to show the star patterns with the least possible distortion and to make it easy to measure positions from the maps. The tick marks along the chart borders and on the central RA line (horizontal RA line on the polar maps) will be helpful.

Stars

Stars are huge balls of hot gas, like our own Sun, but much further away than the Sun. Although they appear as small points of light (so small that even in the largest telescope most of them cannot be measured) many stars are even bigger than our Sun, which turns out to be of only average size. Some stars, like Betelgeuse (in Orion) and Antares (in Scorpius), have diameters that are close to half a billion kilometers (300 times that of our Sun). Even that is not the limit: some stars have diameters a

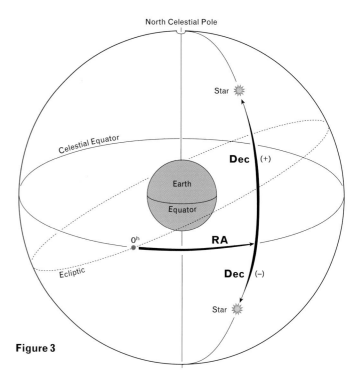

Figure 3

thousand times larger than our Sun. At the other end of the scale are stars that are no bigger than the Earth, and some that are even smaller.

Not only the sizes, but also the temperatures, of stars differ. The temperature of the Sun is almost 5,500 °C. That is the surface temperature; in the nucleus of the Sun the temperature reaches millions of degrees. The Sun is a star that is radiating yellowish light. 'Cooler' stars are orange (like Aldebaran in Taurus, or Arcturus in Boötes, with surface temperatures of 4,000 °C) or red (Betelgeuse and Antares; 3,000 °C). But many stars are much hotter than the Sun. White stars have temperatures between 6,000 and 10,000 °C (Sirius, in Canis Major, or Deneb in Cygnus). Then there are the bluish stars with temperatures between 11,000 and 40,000 °C (like Rigel in Orion). It is interesting to take a look at the beautiful constellation Orion on a bright night in December or January. You can clearly see the difference between the red giant star Betelgeuse and the bluish Rigel. Try it.

Stars are divided into spectral classes, according to differences in their spectra. These differences are related to the surface temperatures and the colors of the stars. The hottest, bluish stars are the classes O, B and A, the slightly cooler white stars are class F, yellow class G, orange class K and the coolest, the red stars, are class M. To remember this rather illogical sequence of letters, use the sentence: **Oh, Be A Fine Girl, Kiss Me!** or if you prefer, **Oh, Be A Fine Guy, Kiss Me!** In the tables, facing the star charts, the spectral classes of variable stars are given in the final column. The spectral types of the 95 brightest stars can be found in Table C on page 35.

Magnitudes

The word magnitude usually refers to the apparent brightness of a star or an object. Traditionally, stars visible to the naked eye were classified into six groups of brightness. The most prominent stars in the sky were called first magnitude stars, the ones slightly fainter (e.g. the Pole Star and the brightest stars of Great Bear) second magnitude and so on. The faintest ones visible to the eye were magnitude six. Nowadays astronomers are able to measure the brightness very accurately. In modern catalogues you will find magnitude to two places of decimals and the scale now has a logarithmic footing. A difference of five magnitudes is defined as 100. Consequently one magnitude represents a difference of 2.5 times (or to be precise 2.512, the fifth root of 100). On this scale several stars turned out to be brighter than 1, so the scale was extended to 0, but even that was not enough. A few stars were still brighter and were given a negative magnitude. The brightest star in the sky, Sirius, has a magnitude of −1.44, the second brightest, Canopus, is of magnitude −0.62.

On the preceding twenty-four monthly sky maps, the brightness of a star is rounded to the nearest whole magnitude. Stars between 0.5 and 1.5 are classified as magnitude 1, between 1.5 and 2.5 magnitude 2 and so on. For the star charts, stars are classified to half magnitudes. A detailed scale can be found in the legend at the right of each chart. Stars'

positions and magnitudes are based on the information from the *Hipparcos and Tycho Catalogues* (ESA Publications, 1997). All together approximately 9,500 stars are plotted. Furthermore, the atlas shows 921 non-stellar objects, i.e. star clusters, nebulae and distant galaxies.

Since stars are at very different distances, the apparent, or visual (V), magnitude does not show how bright the star really is compared with other stars or with our own sun. Astronomers use the term *absolute magnitude* for the real brightness of a star (M_v). The *absolute magnitude* tells you how bright a star would appear in the sky, when it was at a distance of *10 parsecs*. One *parsec* equals a distance of *3.26 light-years*. So 10 parsecs equals 32.6 light-years, or in other words the distance light travels in 32.6 years. That is 9.46×10^{12} (9,460,000,000,000) kilometers. The velocity of light is 299,792 kilometers per second.

In Table C you will find both the *visual* (V) and the *absolute* (M_v) magnitudes and a column telling you the distance in light-years. It is interesting to compare these columns. The brightest star in the sky (Sirius) is not really that bright when you compare its absolute magnitude (1.5) with that of Deneb (−7.5). This star shines about 1,500 times as bright as Sirius.

Although the Sun is not listed in the table, its absolute magnitude (4.8) shows that it is only a very moderate star, fainter than all the stars in the list, although it is obviously the brightest object in the sky, with a visual magnitude of −26.8. It shines so brightly because on the cosmic scale it is very close, 'only' 8 light-minutes! When you compare Deneb with our Sun you will find that it is almost 100,000 times as bright as the Sun.

A lower case letter v after an entry in the visual magnitude column in Table C means that this is a so-called *variable star*. This is explained in section 'Variable stars' on page 38.

Star names

Many of the brighter stars have proper names of Latin, Greek or Arabic origin, such as Regulus, Altair and Betelgeuse. Only the brightest stars are still referred to by these old names. Most stars now bear designations of numbers and Greek letters. The German celestial cartographer Johann Bayer introduced the Greek letters in the beginning of the seventeenth century. In general, the brightest star in a constellation was given the first letter of the Greek alphabet, alpha (α) the second letter, beta (β), was given to the second brightest and so on, although there are many obvious exceptions to this rule. In Table B you will find the Greek letters and their names.

Another way to identify stars is by numbers. In each constellation the stars are numbered in order of RA. These numbers are usually referred to as Flamsteed numbers. Most of the brighter stars have both a Greek letter and a Flamsteed number. On the star charts you will find all Greek letters and, in addition, the Flamsteed numbers for those stars not having a Greek letter. For stars of magnitude 1.5 and brighter, the proper names are also given; this is also done for a few well-known second magnitude stars: Algol, Mira, Castor and Polaris.

Generally, when in text or a table a star is referred to by a Greek letter or by a Flamsteed number, it is followed by the genitive form of the Latin constellation name, or its official abbreviation. So the star Deneb, in the constellation Swan (Latin: *Cygnus)* can also be called Alpha Cygni or 50 Cygni (α Cyg, 50 Cyg). Table D gives the names of the constellations, the genitive form, the official abbreviation and the common English names.

Variable stars (see page 38) have a quite different nomenclature. Some have the regular star identification like Algol (β Per) or Mira (o Cet) but most are labeled in a special way: by roman letters starting with R, then S, T etc. to Z. Then RR, RS, RT, . . . , RZ, next SS, ST, and so on up to ZZ. After these 54 the naming continues with AA, AB, . . . , AZ, then BB, BC and so on again up to QZ, giving a total 334. The next variable found in that constellation is called V335, then V336 and so on. These designations are also followed by the constellation name, as with the Greek letters and the Flamsteed numbers.

Table B   *The Greek alphabet*

α	Alpha	ι	Iota	ρ	Rho
β	Beta	κ	Kappa	σ	Sigma
γ	Gamma	λ	Lambda	τ	Tau
δ	Delta	μ	Mu	υ	Upsilon
ε	Epsilon	ν	Nu	φ	Phi
ζ	Zeta	ξ	Xi	χ	Chi
η	Eta	o	Omicron	ψ	Psi
ϑ	Theta	π	Pi	ω	Omega

Constellations

Most of the constellations we know originate from Mesopotamian traditions and from Greek mythology, but over the centuries many other constellations have been added to the classical ones, especially in the southern sky. Several of these new constellations only had a short life, while others have survived. In 1930 the International Astronomical Union (IUA) finally adopted a list of 88 official constellations and the boundaries were also delimited once and for all. On our charts these official boundaries are drawn in as dashed lines. The final column of Table D gives the numbers of the charts where you can find each of these constellations.

Table C   *The 95 brightest stars*

All stars brighter than magnitude 2.55 (v), in order of brightness

Star	Name	RA h	m	Dec °	′	Magnitudes V	M_v	Spectral type	Distance (light-years)	Chart numbers			
α CMa	Sirius	06	45.2	−16	43	−1.44	1.5	A0	9	9	15		
α Car	Canopus	06	23.9	−52	41	−0.62	−5.4	F0	313	14	15	16	
α Cen	Rigil Kent	14	39.6	−60	50	−0.29	4.0	G2 + K1	4	17	18	20	
α Boo	Arcturus	14	15.7	+19	11	−0.05	−0.6	K2	37	5	11		
α Lyr	Vega	18	36.9	+38	47	0.03	0.6	A0	25	6	7		
α Aur	Capella	05	16.7	+46	00	0.08	−0.8	G6	42	2	3		
β Ori	Rigel	05	14.5	−08	12	0.18	−6.6	B8	773	9			
α CMi	Procyon	07	39.3	+05	14	0.40	2.8	F5	11	9	10		
α Ori	Betelgeuse	05	55.2	+07	24	0.45 v	−5.0	M2	427	9			
α Eri	Achernar	01	37.6	−57	14	0.45	−2.9	B3	144	14	15	19	20
β Cen	Hadar	14	03.8	−60	22	0.61 v	−5.5	B1	525	16	17	18	20
α Aql	Altair	19	50.8	+08	52	0.76	2.1	A7	17	7	12	13	
α Cru	Acrux	12	26.6	−63	06	0.77	−4.6	B0	321	16	17	20	
α Tau	Aldebaran	04	35.9	+16	31	0.87 v	−0.8	K5	65	2	3	8	9
α Vir	Spica	13	25.2	−11	10	0.98 v	−3.6	B1	262	11			
α Sco	Antares	16	29.4	−26	26	1.06 v	−5.8	M1	604	11	12	17	18

Table C *The 95 brightest stars (continued)*

All stars brighter than magnitude 2.55 (v), in order of brightness

Star	Name	RA h	m	Dec °	'	Magnitudes V	M$_v$	Spectral type	Distance (light-years)	Chart numbers			
β Gem	Pollux	07	45.3	+28	02	1.16	1.1	K0	34	3	4	9	10
α PsA	Fomalhaut	22	57.7	−29	37	1.17	1.6	A3	25	14	19		
α Cyg	Deneb	20	41.4	+45	17	1.25	−7.5	A2	3262	6	7		
β Cru	Mimosa	12	47.7	−59	41	1.25 v	−4.0	B0	353	16	17	20	
α Leo	Regulus	10	08.4	+11	58	1.36	−0.6	B7	77	10			
ε CMa	Adhara	06	58.6	−28	58	1.50	−4.1	B2	431	9	15	16	
α Gem	Castor	07	34.6	+31	53	1.58	0.6	A2	52	3	4		
γ Cru	Gacrux	12	31.2	−57	07	1.59 v	−0.8	M4	88	16	17	20	
λ Sco	Shaula	17	33.6	−37	06	1.62 v	−3.6	B1	701	17	18		
γ Ori	Bellatrix	05	25.1	+06	21	1.64	−2.8	B2	243	9			
β Tau	Alnath	05	26.3	+28	36	1.65	−1.3	B7	131	3	9		
β Car	Miaplacides	09	13.2	−69	43	1.67	−1.1	A2	111	16	20		
ε Ori	Alnilam	05	36.2	−01	12	1.69	−6.6	B0	1345	9			
α Gru	Alnair	22	08.2	−46	58	1.73	−0.9	B7	101	14	18	19	20
ζ Ori	Alnitak	05	40.8	−01	57	1.74	−5.5	O9	816	9			
γ Vel	—	08	09.5	−47	20	1.75 v	−5.8	WC8 + O9	842	15	16		
ε UMa	Alioth	12	54.0	+55	58	1.76 v	−0.2	A0	81	1	4	5	
α Per	Mirphak	03	24.3	+49	52	1.79	−4.9	F5	593	2	3		
ε Sgr	Kaus Australis	18	24.2	−34	23	1.79	−1.4	B9	145	18			
α UMa	Dubhe	11	03.7	+61	45	1.81	−1.3	F7	124	1	4	5	
δ CMa	Wezen	07	08.4	−26	23	1.83	−7.2	F8	1787	9	15	16	
η UMa	Alkaid	13	47.5	+49	19	1.85	−1.8	B3	101	1	4	5	6
ε Car	Avior	08	22.5	−59	31	1.86	−4.8	K3 + B2	633	15	16	20	
ϑ Sco	—	17	37.3	−43	00	1.86	−3.0	F1	272	17	18		
β Aur	Menkalinan	05	59.5	+44	57	1.90 v	−0.2	A2	82	2	3	4	
α TrA	Atria	16	48.7	−69	02	1.91	−5.0	K2	415	17	18	20	
δ Vel	—	08	44.7	−54	43	1.93	0.0	A1	80	15	16	20	
γ Gem	Alhena	06	37.7	+16	24	1.93	−0.6	A0	105	3	9		
α Pav	Peacock	20	25.6	−56	44	1.94	−2.1	B2	183	18	19	20	
α UMi	Polaris	02	31.8	+89	16	1.97 v	−4.1	F7	432	1			
β CMa	Mirzam	06	22.7	−17	57	1.98 v	−4.0	B1	500	9	15		
α Hya	Alphard	09	27.6	−08	39	1.99	−2.1	K3	177	10			
γ Leo	Algieba	10	20.0	+19	50	2.01	−2.2	K0	126	4	10		
α Ari	Hamal	02	07.2	+23	28	2.01	0.5	K2	66	2	8		
β Cet	Deneb Kaitos	00	43.6	−17	59	2.04	−1.0	K0	96	8	14		
σ Sgr	Nunki	18	55.3	−26	18	2.05	−2.4	B2.5	224	12	18		
β Gru	—	22	42.7	−46	53	2.06 v	−1.4	M5	170	14	19		
ϑ Cen	Menkent	14	06.7	−36	22	2.06	0.1	K0	61	17			
β UMi	Kochab	14	50.7	+74	09	2.07	−1.1	K4	126	1			
α And	Alpheratz	00	08.4	+29	05	2.07	−0.6	B9	97	2	7	8	13
β And	Mirach	01	09.7	+35	37	2.07	−1.9	M0	199	2	7		
κ Ori	Saiph	05	47.8	−09	40	2.07	−5.0	B0	721	9			
α Oph	Rasalhague	17	34.9	+12	34	2.08	1.3	A5	47	12			
β Per	Algol	03	08.2	+40	57	2.09 v	−0.5	B8	93	2	3		
γ And	Almaak	02	03.9	+42	20	2.10	−3.0	B8	355	2	3	7	
β Leo	Denebola	11	49.1	+14	34	2.14	1.9	A3	36	2	5	10	11
γ Cas	—	00	56.7	+60	43	2.15 v	−5.0	B0	613	1	2	7	
γ Cen	—	12	41.5	−48	58	2.20	−0.6	A1	130	16	17		
ζ Pup	—	08	03.6	−40	00	2.21	−6.1	O5	1403	15	16		
ι Car	Aspidiske	09	17.1	−59	17	2.21 v	−4.4	A8	694	15	16	20	
α CrB	Alphekka	15	34.7	+26	43	2.22 v	0.3	A0	75	5	6	11	12
λ Vel	—	09	08.0	−43	26	2.23 v	−4.8	K4	572	15	16		
γ Cyg	Sadr	20	22.2	+40	15	2.23	−4.1	F8	1517	6	7		
ζ UMa	Mizar	13	23.9	+54	56	2.23	0.3	A2	78	1	4	5	6
γ Dra	Etamin	17	56.6	+51	29	2.24	−1.1	K5	148	5	6	7	
α Cas	Schedir	00	40.5	+56	32	2.24	−2.5	K0	229	1	2	7	
δ Ori	Mintaka	05	32.0	−00	18	2.25 v	−0.6	O9	919	9			
β Cas	Caph	00	09.2	+59	09	2.27 v	1.2	F2	54	1	2	7	
ε Sco	—	16	50.2	−34	18	2.29	0.1	K2	65	17	18		
α Lup	—	14	41.9	−47	23	2.29 v	−4.1	B1	548	17	18	20	
ε Cen	—	13	39.9	−53	28	2.29 v	−3.3	B1	376	16	17	18	20
δ Sco	Alpheratz	16	00.3	−22	37	2.29	−4.4	B0	401	11	12	17	18
η Cen	—	14	35.5	−42	10	2.33 v	−2.8	B1	309	17	18		
β UMa	Merak	11	01.8	+56	23	2.34	0.4	A1	79	1	4	5	
ε Boo	Izar	14	45.0	+27	04	2.35	−2.6	A0	210	5	11		
ε Peg	Enif	21	44.2	+09	52	2.38 v	−5.2	K2	672	13			
κ Sco	—	17	42.5	−39	02	2.39 v	−3.6	B1	464	17	18		
α Phe	Ankaa	00	26.3	−42	18	2.40	−0.3	K0	77	14	19		
γ UMa	Phad	11	53.8	+53	42	2.41	0.2	A0	84	1	4	5	
η Oph	Sabik	17	10.4	−15	43	2.43	0.8	A2	84	12			
β Peg	Scheat	23	03.8	+28	05	2.44 v	−1.7	M2	199	2	7	13	
α Cep	Alderamin	21	18.6	+62	35	2.45	1.4	A7	49	1	6	7	
η CMa	Aludra	07	24.1	−29	18	2.45	−7.5	B5	3182	15	16		
κ Vel	—	09	22.1	−55	00	2.47	−3.9	B2	539	15	16	17	20
ε Cyg	—	20	46.2	+33	58	2.48	0.7	K0	72	6	7		
α Peg	Merkab	23	04.8	+15	12	2.49	−0.9	B9	140	7	13		
α Cet	Menkar	03	02.3	+04	05	2.54	−1.7	M2	220	8			
ζ Cen	—	13	55.5	−47	17	2.54	−2.9	B2	385	16	17	18	20
ζ Oph	—	16	37.2	−10	34	2.54	−4.3	O9	458	11	12		

Table D *List of constellations*

Name	Genitive	Abbreviation	Common name	Chart number(s)		
Andromeda	Andromedae	And	Andromeda	2		
Antlia	Antliae	Ant	Air Pump	16		
Apus	Apodis	Aps	Bird of Paradise	20		
Aquarius	Aquarii	Aqr	Water Carrier	13		
Aquila	Aquilae	Aql	Eagle	12		
Ara	Arae	Ara	Altar	18		
Aries	Arietis	Ari	Ram	8	2	
Auriga	Aurigae	Aur	Charioteer	3		
Boötes	Boötis	Boo	Herdsman	5	11	
Caelum	Caeli	Cae	Engraving Tool	15		
Camelopardalis	Camelopardalis	Cam	Giraffe	1	3	
Cancer	Cancri	Cnc	Crab	10	4	
Canes Venatici	Canum Venaticorum	CVn	Hunting Dogs	5		
Canis Major	Canis Majoris	CMa	Greater Dog	9	15	
Canis Minor	Canis Minoris	CMi	Lesser Dog	9		
Capricornus	Capricorni	Cap	Sea Goat	13		
Carina	Carinae	Car	Keel	16	20	
Cassiopeia	Cassiopeiae	Cas	Cassiopeia	1	2	
Centaurus	Centauri	Cen	Centaur	17		
Cepheus	Cephei	Cep	Cepheus	1		
Cetus	Ceti	Cet	Whale	8		
Chamaeleon	Chamaeleonis	Cha	Chameleon	20		
Circinus	Circini	Cir	Pair of Compasses	17	20	
Columba	Columbae	Col	Dove	15		
Coma Berenices	Coma Berenicis	Com	Berenice's Hair	5	11	
Corona Australis	Coronae Australis	CrA	Southern Crown	18		
Corona Borealis	Coronae Borealis	CrB	Northern Crown	5	6	
Corvus	Corvi	Crv	Crow	11		
Crater	Crateris	Crt	Cup	10		
Crux	Crucis	Cru	Southern Cross	16	17	20
Cygnus	Cygni	Cyg	Swan	7		
Delphinus	Delphini	Del	Dolphin	13		
Dorado	Doradus	Dor	Goldfish	15		
Draco	Draconis	Dra	Dragon	1	6	
Equuleus	Equulei	Equ	Little Horse	13		
Eridanus	Eridani	Eri	River Eridanus	8	9	14
Fornax	Fornacis	For	Furnace	14		
Gemini	Geminorum	Gem	Twins	3	9	
Grus	Gruis	Gru	Crane	19		
Hercules	Herculis	Her	Hercules	6	12	
Horologium	Horologii	Hor	Pendulum Clock	14		
Hydra	Hydrae	Hya	Water Snake	10	16	17
Hydrus	Hydri	Hyi	Lesser Water Snake	20		
Indus	Indi	Ind	Indian	19	20	

Table D *List of constellations (continued)*

Name	Genitive	Abbreviation	Common name	Chart number(s)	
Lacerta	Lacertae	Lac	*Lizard*	7	
Leo	Leonis	Leo	*Lion*	10	4
Leo Minor	Leonis Minoris	LMi	*Lesser Lion*	4	
Lepus	Leporis	Lep	*Hare*	9	
Libra	Librae	Lib	*Scales*	11	17
Lupus	Lupi	Lup	*Wolf*	17	18
Lynx	Lyncis	Lyn	*Lynx*	4	
Lyra	Lyrae	Lyr	*Lyre*	6	
Mensa	Mensae	Men	*Table Mountain*	20	
Microscopium	Microscopii	Mic	*Microscope*	19	
Monoceros	Monocerotis	Mon	*Unicorn*	9	
Musca	Muscae	Mus	*Fly*	20	
Norma	Normae	Nor	*Level*	17	18
Octans	Octantis	Oct	*Octant*	20	
Ophiuchus	Ophiuchi	Oph	*Serpent Holder*	12	18
Orion	Orionis	Ori	*Orion, the Hunter*	9	
Pavo	Pavonis	Pav	*Peacock*	20	18
Pegasus	Pegasi	Peg	*Pegasus*	13	7
Perseus	Persei	Per	*Perseus*	2	3
Phoenix	Phoenicis	Phe	*Phoenix*	14	
Pictor	Pictoris	Pic	*Painter's Easel*	15	
Pisces	Piscium	Psc	*Fishes*	8	13 2
Piscis Austrinus	Piscis Austrini	PsA	*Southern Fish*	19	
Puppis	Puppis	Pup	*Stern*	15	9
Pyxis	Pyxidis	Pyx	*Mariner's Compass*	16	
Reticulum	Reticuli	Ret	*Net*	14	15
Sagitta	Sagittae	Sge	*Arrow*	6	12
Sagittarius	Sagittarii	Sgr	*Archer*	18	12
Scorpius	Scorpii	Sco	*Scorpion*	18	12
Sculptor	Sculptoris	Scl	*Sculptor*	14	19
Scutum	Scuti	Sct	*Shield*	12	
Serpens	Serpentis	Ser	*Serpent*	12	11
Sextans	Sextantis	Sex	*Sextant*	10	
Taurus	Tauri	Tau	*Bull*	9	3
Telescopium	Telescopii	Tel	*Telescope*	18	19
Triangulum	Trianguli	Tri	*Triangle*	2	
Triangulum Australe	Trianguli Australis	TrA	*Southern Triangle*	20	
Tucana	Tucanae	Tuc	*Toucan*	20	
Ursa Major	Ursae Majoris	UMa	*Great Bear*	4	1
Ursa Minor	Ursae Minoris	UMi	*Lesser Bear*	1	
Vela	Velorum	Vel	*Sail*	16	
Virgo	Virginis	Vir	*Virgin*	11	
Volans	Volantis	Vol	*Flying Fish*	20	
Vulpecula	Vulpeculae	Vul	*Fox*	6	7

Variable stars

The brightness of many stars varies over longer or shorter periods of time. The most common reason for this is that the size of the star actually changes: the star pulsates. A well-known type of pulsating star is the Cepheid, named after Delta (δ) Cephei, a yellow supergiant, which regularly pulsates every few days or weeks. The Cepheids are divided into two classes: the classical (Cep) and the Population II (CW) Cepheids. They are important to astronomers because there is a relation between their period and luminosity. The brighter a Cepheid, the longer the period. When the period is measured, we know the real luminosity (*absolute* brightness) of the star. By comparing this with the amount of light we actually receive (apparent or *visual* brightness) we have an important tool for calculating distance.

Another type of pulsating variable is named after the prototype Omicron (o) Ceti or Mira (M), a red giant. These variables do not have a strict period. Several other types of pulsating variables are also named after prototypes, like U Gem, R CrB, or RR Lyr. There is also a completely different type of variable star: the eclipsing variable. An eclipsing variable star is a double star in mutual orbit and one component periodically moves in front or behind the other, causing a drop in the total amount of light we receive. The best-known example of this type is Beta (β) Persei (Algol). Eclipsing variables are referred to in the star chart tables as type E, subdivided by the shape of their light curves into E, EA, EB and EW.

A further group is termed the eruptive variables. These undergo a very sudden and very large increase in brightness. The best known of these are the novae (N) and the supernovae (SN). A nova is a very close double star, in which one component is a white dwarf: a small but very compact star. Gas from the other component flows into the white dwarf and eventually it ignites in a huge explosion. The brightness of the star increases temporarily by thousands of times. Some novae erupt more than once. These are known as recurrent novae (Rn).

A supernova is even more spectacular. It is the catastrophic death of a very hot star. The star's life ends when it blows itself up and for a short period it shines millions of times brighter than it was. After the star has faded again the outer shells of the star form a slowly expanding nebula. The *Crab Nebula* (M1) in Taurus (the Bull) is one example. In the star charts all variable stars with a maximum of magnitude 6.5 or brighter are plotted.

Double stars

The majority of stars are double or multiple stars. Sometimes two stars appear very close in the sky, but are only on the same line of sight; their distances differ considerably. These are called optical double stars. Real physical double stars belong together and are also called binaries. They are tied together by gravity and are in mutual orbit. The same goes for triple, quadruple and even larger families of stars. Their apparent separation is measured in minutes and seconds of arc. One degree (°) equals 60 minutes (′) and one minute again equals 60 seconds (″). The separation is given in the tables (column with heading Sep) in seconds of arc. There is also a column with the heading PA, meaning position angle. This gives the angular position of the fainter component in relation to the brighter one. The angle is measured from the north, eastward. Keep in mind that in the sky east and west are reversed. So, when north is up, east is to the left! Consequently, the position angle is measured anti-clockwise.

All double stars with an integrated (combined) magnitude of 6.5 or brighter are plotted on the charts. The star chart tables contain only a selection of the finest targets.

Open and globular clusters

Open and globular star clusters appear on the star charts as yellow disks. Open clusters are shown with a dotted outline and globular clusters are shown as solid circles with a cross through the center. Open clusters are usually found near the plane of the Milky Way, and so are in or close to the soft blue areas on the charts, representing the brightest parts of the Milky Way. Open clusters are groups of young stars, often hot and bluish, and their individual stars can easily be seen with a small telescope or sometimes with the naked eye. The last column of this section of the star chart table (N*) gives the approximate number of stars in the cluster.

Globular clusters are quite different. They contain larger numbers of stars and are much more compact. They are found outside the galaxy plane and the stars are older. All clusters down to magnitude 10 are plotted on the star charts. The diameter (Diam) in the star chart table is given in minutes (') of arc.

In the first column of the star chart tables you will find the designation of the cluster. First the NGC or IC numbers are listed (NGC New General Catalogue; IC Index Catalogue). NGC numbers are without a prefix (both on the charts and in the tables), and IC numbers have the prefix 'I.'. Alternative names, as well as Messier numbers (M), are in the second column. The same goes for nebulae and galaxies.

Diffuse and planetary nebulae

Diffuse nebulae are areas of the raw materials, dust and gas, from which stars are born. These diffuse nebulae are also found along the spiral arms of the Milky Way, and are visible in other nearby galaxies as well. There are three general types of diffuse nebulae. In an *emission nebulae* (E) one or more hot stars causes the nebula to emit light of its own. A *reflection nebula* (R) only reflects light from nearby stars. Reflection nebulae show a blue color, in contrast to the reddish glow of emission nebulae. The third type is the dark nebula. Most dark nebulae are not easily found in amateur telescopes. Only one is shown on the maps: the Coal Sack in the constellation Crux (Southern Cross), a dark patch in the Milky Way and visible with the naked eye (Charts 16, 17 and 20).

Planetary nebulae have nothing to do with planets. The name arose from their disk-like appearance. They are almost spherical cast-off shells of gas from very hot stars, late in their life spans. Often the ionized gas has a greenish color. Examples are the Ring Nebula (M57) in Lyra (the Lyre) and the Helix Nebula (NGC 7293) in Aquarius (the Water Carrier).

Both bright diffuse nebulae and planetary nebulae are printed in soft green on the atlas charts.

Galaxies

The red ovals on the charts are the most remote objects: the galaxies. Galaxies are huge systems of stars, clusters and nebulae, like our own Milky Way. There are several types of galaxy: the elliptical (E), spiral (S), barred spiral (SB) and irregular. The E type is subdivided according to shape: from E0 for the almost spherical to E7 for the flattened lens shape. The S and SB types are subdivided according to how tightly the spiral arms are wound (See Figure 4).

The Galactic Equator

On the charts the Galactic Equator is drawn as a line of dots and dashes. It represents the projection of the plane of our galaxy on the stellar sphere. Every ten degrees of galactic longitude is marked along this equator. The center of our Milky Way is at 0° galactic longitude, and can be found on chart 18, in Sagittarius, close to the Ophiuchus boundary.

The Ecliptic

The Ecliptic is the projection of the Earth's orbit around the Sun, or the yearly path of the Sun along the sky, caused by the orbital movement of the Earth. It is shown on the charts as a broken line. Every ten degrees of longitude is marked. The ecliptic longitude starts at the vernal equinox, at right ascension 0^h and declination 0° (Charts 8 and 13). The vernal equinox is where the Sun is when it is exactly in the plane of the Earth's Equator, around March 21 each year. The Moon and the planets are always close to the Ecliptic.

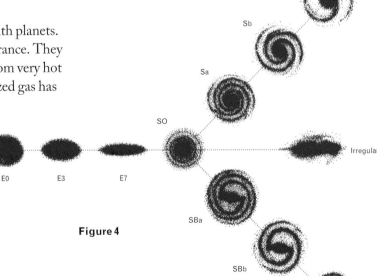

Figure 4

Abbreviations used in the star chart tables

A table, giving information of all the interesting telescopic objects on the charts accompanies each star chart. The listing is restricted to the main regions of each chart excluding areas of overlap. These regions are indicated in each table heading and are shown on the Index to the star charts on the facing page (41). The constellation abbreviations follow the official IAU practice (see the list of constellations in Table D).

For the double stars only a selection is shown, limited to double stars which can be separated in small or medium size telescopes. As far as possible all plotted variable stars, open and globular clusters, planetary and diffuse nebulae, and galaxies are listed. All positions are for the epoch 2000.0.

In general visual (V) magnitudes are listed, except for the planetary nebulae, where the photographic magnitudes are given. Elsewhere, when a photographic magnitude is used, the figure is followed by 'p'.

When an entry in the tables is followed by a colon, it means that the given value is uncertain.

General

Con	Constellation (the official IAU abbreviations: see Table D)
RA	Right ascension
Dec	Declination
NGC/IC	Numbers from the New General Catalogue (NGC) or Index Catalogue (IC) by J.L.E. Dryer. NEC numbers have no prefix, IC numbers have the prefix I

Variable stars

Range (max – min), type, period (days) and spectrum are given. In general only variable stars with a range of more than 0.5 magnitude are plotted and listed.

Type:

Cep	Classical Cepheid
CW	Type II Cepheid
E	Eclipsing binary
EA	Algol type
EB	Beta Lyrae type
EW	W Ursae Majoris type
M	Mira (long-period) type
SR	Semi-regular
Irr	Irregular
RCB	R Coronae Borealis type
δ Sct	Delta Scuti type
ZA	Z Andromedae type
Rn	Recurrent nova
N	Nova
RV	RV Tauri type
SD	S Dor type

Double stars

In general double stars with a separation less than 1 minute of arc, or with components fainter than magnitude 10.0, are not listed.

When a magnitude entry is followed by 'd', it means that this component itself is also a double star.

PA	Position angle, from north (0°) through east (90°), south (180°) and west (270°)
Sep	Separation, in seconds of arc. The data are for the year 2000

Open clusters

Mag	Integrated visual magnitude
Diam	Diameter, in minutes of arc
N*	Approximate number of stars

Globular clusters

Diam	Diameter, in minutes of arc
Mag	Integrated visual magnitude

Bright diffuse nebulae

Type	Type of nebulae E Emission nebula R Reflection nebula
Diam	Angular dimensions, in minutes of arc
Mag*	Approximate magnitude of the illuminating star

Planetary nebulae

Diam	Diameter, in *seconds* of arc Two values, separated by a slash (/) refer to a bright core, surrounded by a fainter halo
Mag	Integrated photographic magnitude
Mag*	Photoelectric magnitude of the central star

Galaxies

Mag	Integrated visual magnitude
Diam	Major and minor diameters, in minutes of arc
Type	Type of galaxy (see Figure 4 on page 39)

Index to the star charts

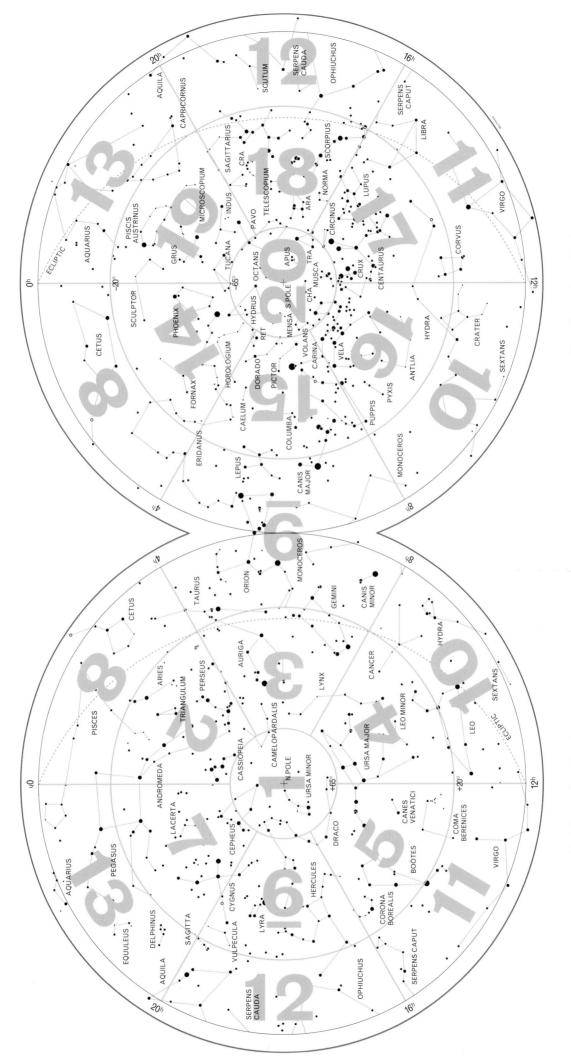

Southern hemisphere

Northern hemisphere

Chart 1 *North of declination +65°*

Variable stars

Star	Con	RA h m	Dec ° '	Range	Type	Period (days)	Spectrum
YZ	Cas	00 45.7	+74 59	5.7–6.1	EA	4.47	A + F
RZ	Cas	02 48.9	+69 38	6.2–7.7	EA	1.20	A
R	UMa	10 44.6	+68 47	6.7–13.4	M	301.7	M
VY	UMa	10 45.1	+67 25	5.9–6.5	Irr	—	M
RY	Dra	12 56.4	+66 00	6.0–8.0	SR	172.5:	K
RR	UMi	14 57.6	+65 56	6.1–6.5 p	SR	40:	M
UX	Dra	19 21.6	+76 34	5.9–7.1	SR	168:	C

Double stars

Star	Con	RA h m	Dec ° '	PA °	Sep "	Magnitudes	
ψ	Cas	01 25.9	+68 08	025	33.4	4.7 + 9.6	
48	Cas	02 02.0	+70 54	263	0.9	4.7 + 6.4	Binary, 60.4 years
ι	Cas	02 29.1	+67 24	230	2.5	4.6 + 6.9	Binary, 840 years
				114	7.2	8.4	
α	UMi	02 31.8	+89 16	218	18.4	2.0 + 9.0	Polaris
ψ	Dra	17 41.9	+72 09	015	30.3	4.9 + 6.1	
40,41	Dra	18 00.2	+80 00	232	19.3	5.7 + 6.1	
ε	Dra	19 48.2	+70 16	015	3.1	3.8 + 7.4	
κ	Cep	20 08.9	+77 43	122	7.4	4.4 + 8.4	
β	Cep	21 28.7	+70 34	249	13.3	3.2 + 7.9	Alfirk
π¹	Cep	23 07.9	+75 23	357	1.2	4.6 + 6.6	Binary, 147 years
ο	Cep	23 18.6	+68 07	223	2.8	4.9 + 7.1	Binary, 796.2 years

Open cluster

NGC/IC	Other	Con	RA h m	Dec ° '	Diam '	Mag	N*
188		Cep	00 44.4	+85 20	14	8.1	120

Bright diffuse nebulae

NGC/IC	Other	Con	RA h m	Dec ° '	Diam '	Type	Mag*
7822		Cep	00 03.6	+68 37	60×30	E	
—	Ced 214	Cep	00 04.7	+67 10	50×40	E + R	5.7
7023		Cep	21 01.8	+68 12	18×18	R	6.8

Planetary nebulae

NGC/IC	Other	Con	RA h m	Dec ° '	Mag	Diam "	Mag*
40		Cep	00 13.0	+72 32	10.7	37	11.6
I.3568		Cam	12 32.9	+82 33	11.6	6	12.3
6543		Dra	17 58.6	+66 38	8.8	18/350	11.4

Galaxies

NGC/IC	Other	Con	RA h m	Dec ° '	Mag	Size	Type
I.342		Cam	03 46.8	+68 06	9.1	17.8×17.4	Sc
I.356		Cam	04 07.8	+69 49	11.4	5.2×4.1	Sb
1560		Cam	04 32.8	+71 53	11.5	9.8×2.0	Sd
1961		Cam	05 42.1	+69 23	11.1	4.3×3.0	Sb
2146		Cam	06 18.7	+78 21	10.5	6.0×3.8	SBb
2336		Cam	07 27.1	+80 11	10.5	6.9×4.0	Sb
2366		Cam	07 28.9	+69 13	10.9	7.6×3.5	Irr
2403		Cam	07 36.9	+65 36	8.4	17.8×11.0	Sc
—	U4305	UMa	08 18.9	+70 43	10.6	7.6×6.2	Irr
2655		Cam	08 55.6	+78 13	10.1	5.1×4.1	SBa
2715		Cam	09 08.1	+78 05	11.4	5.0×1.9	Sc
2787		UMa	09 19.3	+69 12	10.8	3.4×2.3	Sa
2976		UMa	09 47.3	+67 55	10.2	4.9×2.5	Sc
2985		UMa	09 50.4	+72 17	10.5	4.3×3.4	Sb
3031	M81	UMa	09 55.6	+69 04	6.9	25.7×14.1	Sb
3034	M82	UMa	09 55.8	+69 41	8.4	11.2×4.6	Pec
3077		UMa	10 03.3	+68 44	9.9	4.6×3.6	E2
3147		Dra	10 16.9	+73 24	10.7	4.0×3.5	Sb
I.2574		UMa	10 28.4	+68 25	10.6	12.3×5.9	S
3348		UMa	10 47.2	+72 50	11.2	2.2×2.2	E1
4125		Dra	12 08.1	+65 11	9.8	5.1×3.2	E5
4236		Dra	12 16.7	+69 28	9.7	18.6×6.9	SB
4589		Dra	12 37.4	+74 12	11.8	3.0×2.7	Sa
4750		Dra	12 50.1	+72 52	11.9	2.3×2.1	Sa
6503		Dra	17 49.4	+70 09	10.2	6.2×2.3	Sb

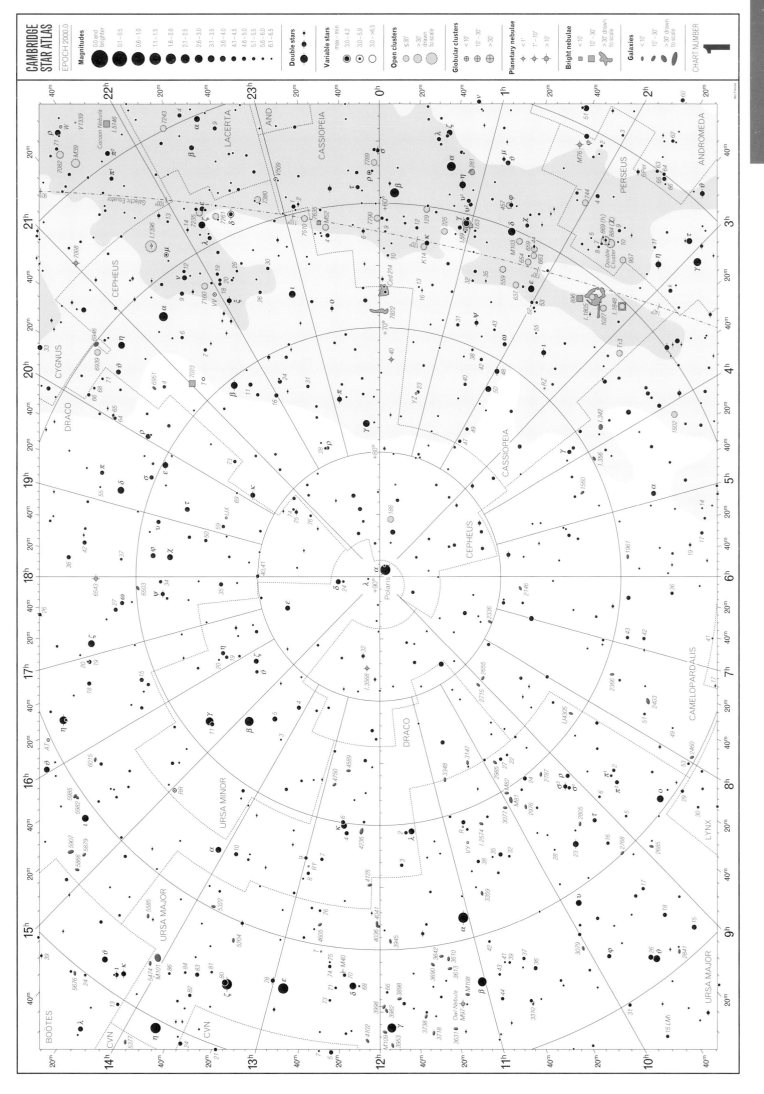

Chart 2 RA 0ʰ to 4ʰ, declination +65° to +20°

Variable stars

Star	Con	RA h m	Dec ° '	Range	Type	Period (days)	Spectrum
R	And	00 24.0	+38 35	5.8–14.9	M	409.3	M
γ	Cas	00 56.7	+60 43	1.6–3.0	Irr	—	B
R	Tri	02 37.0	+34 16	5.4–12.6	M	266.5	M
ρ	Per	03 05.2	+38 50	3.3–4.0	EA	50:	M
β	Per	03 08.2	+40 57	2.1–3.4	EA	2.87	B+G Algol
X	Per	03 55.4	+31 03	6.1–7.0	Irr	—	O

Double stars

Star	Con	RA h m	Dec ° '	PA °	Sep "	Magnitudes	
π	And	00 36.9	+33 43	173	35.9	4.4+8.6	
η	Cas	00 49.1	+57 49	317	12.9	3.4+7.5	Binary, 480 years
γ	Cas	00 56.7	+60 43	248	2.1	1.6v+11.2	Variable
ψ¹	Psc	01 05.6	+21 28	159	30.0	5.6+5.8	
35	Cas	01 21.1	+64 40	344	55.5	6.3+8.7	
γ	And	02 03.9	+42 20	63	9.8	2.3+5.1 d	Almaak
				103	0.4	5.5+6.3	Binary, 61.1 years
6	Tri	02 12.4	+30 18	71	3.9	5.3+6.9	= ε Tri
30	Ari	02 37.0	+24 39	274	38.6	6.6+7.4	
33	Ari	02 40.7	+27 04	0	28.6	5.5+8.4	
ε	Ari	02 59.2	+21 20	203	1.4	5.2+5.5	
ϑ	Per	02 44.2	+49 14	305	20.0	4.1+9.9	Binary, 2720 years
η	Per	02 50.7	+55 54	300	28.3	3.8+8.5	
				268	66.6	9.8	
40	Per	03 42.4	+33 58	238	20.0	5.0+9.5	
o	Per	03 44.3	+32 17	37	10	3.8+8.3	Atik
ζ	Per	03 54.1	+31 53	208	12.9	2.9+9.5	
				286	32.8	11.3	
				195	94.2	9.5	
				185	120.3	10.2	
ε	Per	03 57.9	+40 01	10	8.8	2.9+8.1	

Open clusters

NGC/IC	Other	Con	RA h m	Dec ° '	Mag	Diam '	N*	
129		Cas	00 29.9	+60 14	6.5	21	35	
—	K14	Cas	00 31.9	+63 10	8.5	7	20	
225		Cas	00 43.4	+61 47	7.0	12	15	
457		Cas	01 19.1	+58 20	6.4	13	80	
559		Cas	01 29.5	+63 18	9.5	4.4	60	
581	M103	Cas	01 33.2	+60 42	7.4	6	25	
637		Cas	01 42.9	+64 00	8.2	3.5	20	
654		Cas	01 44.1	+61 53	6.5	5	60	
659		Cas	01 44.2	+60 42	7.9	5	40	
663		Cas	01 46.0	+61 15	7.1	16	80	
752		And	01 57.8	+37 41	5.7	50	60	
744		Per	01 58.4	+55 29	7.9	11	20	
869	h	Per	02 19.0	+57 09	4.3 p	30	200	Double Cluster
884	χ	Per	02 22.4	+57 07	4.4 p	30	150	Double Cluster
I.1805		Cas	02 32.7	+61 27	6.5	22	40	In nebula
957		Per	02 33.6	+57 32	7.6	11	30	
1039	M34	Per	02 42.0	+42 47	5.2	35	60	
1027		Cas	02 42.7	+61 33	6.7	20	40	
I.1848		Cas	02 51.2	+60 26	6.5	12	10	
—	Tr 3	Cas	03 11.8	+63 15	7.0p	23	30	
1245		Per	03 14.7	+47 15	8.4	10	200	
1342		Per	03 31.6	+37 20	6.7	14	40	
—	M45	Tau	03 47.0	+24 07	1.2	110	100	Pleiades
1444		Per	03 49.4	+52 40	6.6	4.0		

Bright diffuse nebulae

NGC/IC	Other	Con	RA h m	Dec ° '	Type	Diam '	Mag*	
281		Cas	00 52.8	+56 36	E	35×30	7.8	
I.59		Cas	00 56.7	+61 04	E+R	10×5	2.5	γ Cas
I.63		Cas	00 59.5	+60 49	E+R	10×3	2.5	γ Cas
896		Cas	02 24.8	+61 54	E	27×13	10.5	
I.1805		Cas	02 33.4	+61 26	E	60×60		
I.1848		Cas	02 51.3	+60 25	E	60×30		

Planetary nebula

NGC/IC	Other	Con	RA h m	Dec ° '	Mag p	Diam "	Mag*	
650, 651	M76	Per	01 42.4	+51 34	12.2	65/290	17.0:	Little Dumbbell

Galaxies

NGC/IC	Other	Con	RA h m	Dec ° '	Mag	Size	Type	
23		Peg	00 09.9	+25 55	11.9	2.3×1.6	Sc	
147		Cas	00 33.2	+48 30	9.3	12.9×8.1	dE4	
185		Cas	00 39.0	+48 20	9.2	11.5×9.8	dE0	
205	M110	And	00 40.4	+41 41	8.0	17.4×9.8	E6	Companion to M31
221	M32	And	00 42.7	+40 52	8.2	7.6×5.8	E2	Companion to M31
224	M31	And	00 42.7	+41 16	3.5	178×63	Sb	Andromeda Galaxy
278		Cas	00 52.1	+47 33	10.9	2.2×2.1	E0	
404		And	01 09.4	+35 43	10.1	4.4×4.2	E0	Appr. 6.5' NW of β And
598	M33	Tri	01 33.9	+30 39	5.7	62×39	Sc	Triangulum Galaxy
I.1727		Tri	01 47.5	+27 20	11.6	6.2×2.9	SB	
672		Tri	01 47.9	+27 26	10.8	6.6×2.7	SBc	
784		Tri	02 01.3	+28 50	11.8	6.2×1.7	SB	
891		And	02 22.6	+42 21	10.0	13.5×2.8	Sb	
925		Tri	02 27.3	+33 35	10.0	9.8×6.0	SBc	
I.239		And	02 36.5	+38 58	11.2	4.6×4.3	SBc	
1003		Per	02 39.3	+40 52	11.5	5.4×2.1	Sc	
1023		Per	02 40.4	+39 04	9.5	8.7×3.3	E7	

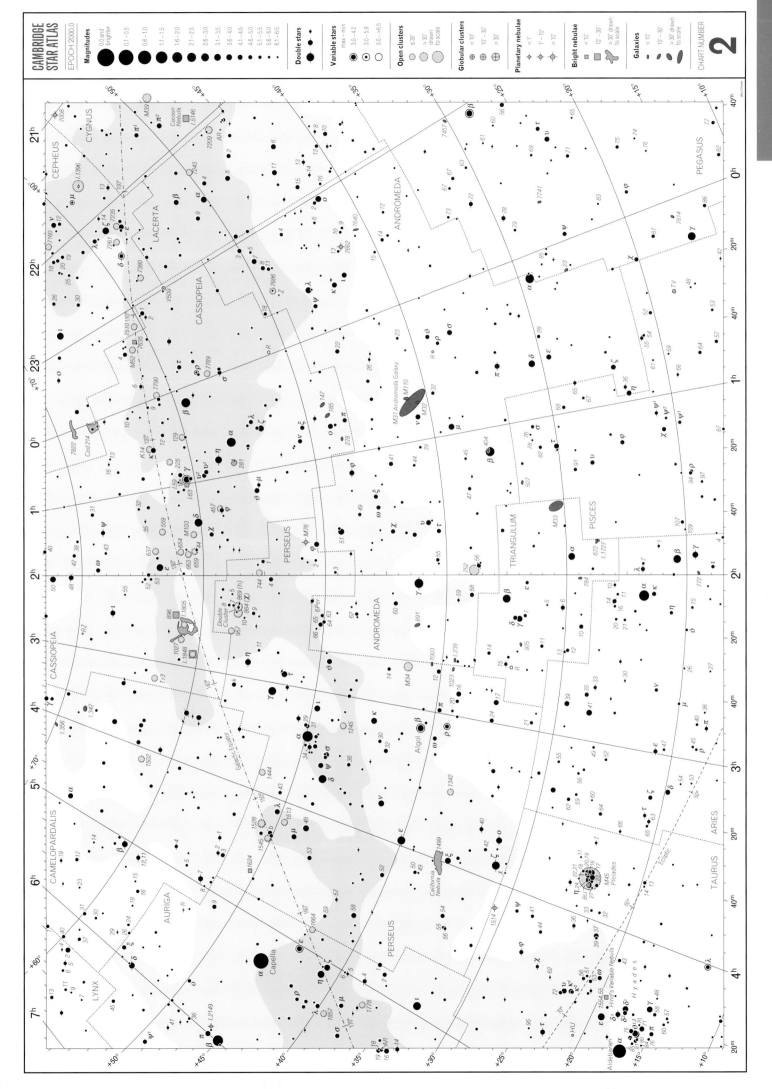

CAMBRIDGE STAR ATLAS

EPOCH 2000.0

Magnitudes
0.0 and brighter
0.1–0.5
0.6–1.0
1.1–1.5
1.6–2.0
2.1–2.5
2.6–3.0
3.1–3.5
3.6–4.0
4.1–4.5
4.6–5.0
5.1–5.5
5.6–6.0
6.1–6.5

Double stars

Variable stars
max – min
3.0–4.2
3.0–5.9
>6.5

Open clusters
≤30'
>30' drawn
drawn to scale

Globular clusters
<10'
10'–30'
>30'

Planetary nebulae
<1'
1'–10'
>10'

Bright nebulae
10'–30'
>30' drawn
to scale

Galaxies
<10'
10'–30'
>30' drawn to scale

CHART NUMBER
2

Wil Tirion

Chart 3 RA 4ʰ to 8ʰ, declination +65° to +20°

Variable stars

Star	Con	RA h m	Dec ° '	Range	Type	Period (days)	Spectrum
HU	Tau	04 38.3	+20 41	5.9–6.7	EA	2.06	A
ε	Aur	05 02.0	+43 49	2.9–3.8	EA	9892	A–F
R	Aur	05 17.3	+53 35	6.7–13.7	M	457.5	M
AR	Aur	05 18.3	+33 46	6.2–6.8	EA	4.13	A+B
U	Ori	05 55.8	+20 10	4.8–12.6	M	372.4	M
TV	Gem	05 11.8	+21 52	8.7–9.5p	SR	182	M
BU	Gem	06 12.3	+22 54	5.7–7.5	Irr	—	M
η	Gem	06 14.9	+22 30	3.2–3.9	SR	232.9	M Propus
RT	Aur	06 28.6	+30 30	5.0–5.8	Cep	3.72	F–G
WW	Aur	06 32.5	+32 27	5.8–6.5	EA	2.53	A
UU	Aur	06 36.5	+38 27	7.8–10.0	SR	234	C
ζ	Gem	07 04.1	+20 34	3.7–4.2	Cep	10.15	F–G Mekbuda
R	Gem	07 07.4	+22 42	6.0–14.0	M	369.8	S

Double stars

Star	Con	RA h m	Dec ° '	PA °	Sep "	Magnitudes	
φ	Tau	04 20.4	+27 21	250	52.1	5.0+8.0	
χ	Tau	04 22.6	+25 38	24	19.4	5.5+7.6	
56	Per	04 24.6	+33 58	22	4.2	5.9+8.7	
κ¹, κ²	Tau	04 25.4	+22 18	173	339	4.2+5.3	
4	Aur	04 59.3	+37 53	359	5.4	5.0+8.0	
14	Aur	05 15.4	+32 41	352	11.1	5.1+11.1	
				226	14.6	7.4	
					184.0	10.4	
R	Aur	05 17.3	+53 35	321	47.5	6.9v+8.6	Variable
118	Tau	05 29.3	+25 09	339	4.8	5.8+6.6	
26	Aur	05 38.6	+30 30	204	0.2	6.0+6.3	Binary, 53.2 years
				267	12.4	8.0	
ν	Aur	05 51.5	+39 09	206	54.6	4.0+9.3	
δ	Aur	05 59.5	+54 17	271	115.4	3.7+9.5	
				67	197.1	9.5	
θ	Aur	05 59.7	+37 13	313	3.6	2.6+7.1	
				297	50.0	10.6	
η	Gem	06 14.9	+22 30	257	1.6	3.3v+8.8	Propus; var; bin, 473.7 y.
μ	Gem	06 22.9	+22 31	141	121.7	3.2+9.4	
ν	Gem	06 29.0	+20 13	329	112.5	4.2+8.7	
ε	Gem	06 43.9	+25 08	94	110.3	3.0+9.0	Mebsuta
12	Lyn	06 46.2	+59 27	70	1.7	5.4+6.0	Binary, 699 years
				308	8.7	7.3	
ζ	Gem	07 04.1	+20 34	84	87.0	3.7v+10.5	Variable
				350	96.5	8.0	
δ	Gem	07 20.1	+21 59	226	5.8	3.5+8.2	Wasat; binary, 1,200 years
19	Lyn	07 22.9	+55 17	315	14.8	5.6+6.5	
				3	214.9	8.9	
α	Gem	07 34.6	+31 53	68	4.0	1.9+2.9	Castor; binary, 420 years
				163	72.5	8.8	
24	Lyn	07 43.0	+58 43	320	54.7	5.0+9.5	

Open clusters

NGC/IC	Other	Con	RA h m	Dec ° '	Mag	Diam '	N*	
1502		Cam	04 07.7	+62 20	5.7	8	45	
1513		Per	04 10.0	+49 31	8.4	9	50	
1528		Per	04 15.4	+51 14	6.4	24	40	
1545		Per	04 20.9	+50 15	6.2	18	20	
1664		Aur	04 51.1	+43 42	7.6	18		
1746		Tau	05 03.6	+23 49	6.1	42	20	
1778		Aur	05 08.1	+37 03	7.7	7	25	
1857		Aur	05 20.2	+39 21	7.0	6	40	
1893		Aur	05 22.7	+33 24	7.5	11	60	
1907		Aur	05 28.0	+35 19	8.2	7	30	
1912	M38	Aur	05 28.7	+35 50	6.4	21	100	
1960	M36	Aur	05 36.1	+34 08	6.0	12	60	
2099	M37	Aur	05 52.4	+32 33	5.6	24	150	
2129		Gem	06 01.0	+23 18	6.7	7	40	
I.2157		Gem	06 05.0	+24 00	8.4	7	20	
2158		Gem	06 07.5	+24 06	8.6	5		
2168	M35	Gem	06 08.9	+24 20	5.1	28	200	
2175		Ori	06 09.8	+20 19	6.8	18	60	In nebula 2174
2281		Aur	06 49.3	+41 04	5.4	15	30	
2331		Gem	07 07.2	+27 21	8.5	18	30	
2420		Gem	07 38.5	+21 34	8.3	10	100	

Globular cluster

NGC/IC	Other	Con	RA h m	Dec ° '	Mag	Diam '
2419		Lyn	07 38.1	+38 53	10.4	4.1

Bright diffuse nebulae

NGC/IC	Other	Con	RA h m	Dec ° '	Type	Diam '	Mag*	
1499		Per	04 00.7	+36 37	E	145×40	4.0	California Nebula
1624		Per	04 40.5	+50 27	E+R	5×5		
1931		Aur	05 31.4	+34 15	E+R	3×3		
1952	M1	Tau	05 34.5	+22 01	E	6×4	16	Crab Nebula; SNR
2174		Gem	06 09.7	+20 30	E	40×30	7.6	Contains cluster 2175
I.443		Gem	06 16.9	+22 47	E	50×40	8.8	Supernova remnant

Planetary nebulae

NGC/IC	Other	Con	RA h m	Dec ° '	Mag p	Diam "	Mag*	
1514		Tau	04 09.2	+30 47	10.0	114	9.4	
2392		Gem	07 29.2	+20 55	9.9	13/44	10.5	Eskimo Nebula

Galaxy

NGC/IC	Other	Con	RA h m	Dec ° '	Mag	Size '	Type
2460		Cam	07 56.9	+60 21	11.7	2.9×2.2	Sb

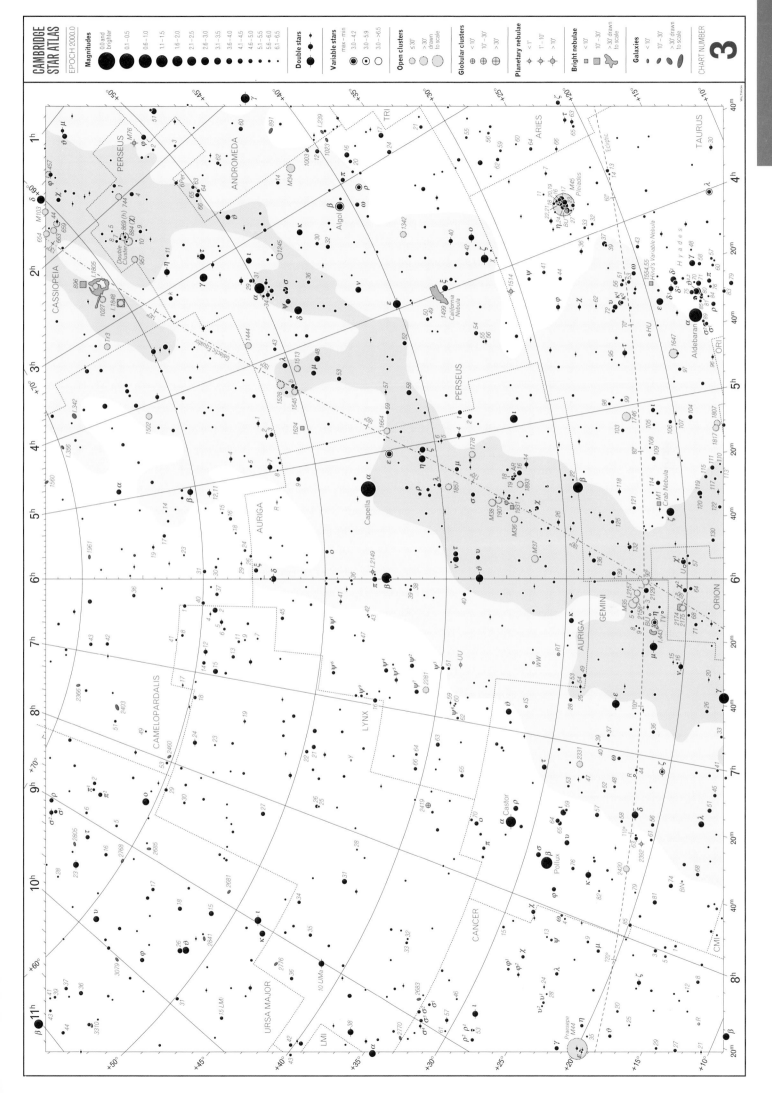

Magnitudes

0.0 and brighter · 0.1–0.5 · 0.6–1.0 · 1.1–1.5 · 1.6–2.0 · 2.1–2.5 · 2.6–3.0 · 3.1–3.5 · 3.6–4.0 · 4.1–4.5 · 4.6–5.0 · 5.1–5.5 · 5.6–6.0 · 6.1–6.5

Double stars

Variable stars

	max—min
	3.0 – 4.2
	3.0 – 5.9
	3.0 – >6.5

Open clusters ≤30′ / >30′ drawn to scale

Globular clusters <10′ / 10′–30′ / >30′

Planetary nebulae <1′ / 1′–10′ / >10′

Bright nebulae <10′ / 10′–30′ / >30′ drawn to scale

Galaxies <10′ / 10′–30′ / >30′ drawn to scale

Star chart 3

Chart 4 RA 8ʰ to 12ʰ, declination +65° to +20°

Variable stars

Star	Con	RA h m	Dec ° '	Range	Type	Period (days)	Spectrum
RS	Cnc	09 10.6	+30 58	5.1–7.0	SR	120:	M
R	LMi	09 45.6	+34 31	6.3–13.2	M	371.9	M
ST	UMa	11 27.8	+45 11	7.7–9.5	SR	81	M

Double stars

Star	Con	RA h m	Dec ° '	PA °	Sep "	Magnitudes	Spectrum
ι	Cnc	08 46.7	+28 46	307	30.5	4.2 + 6.6	
ι	UMa	08 59.2	+48 02	177	2.0	3.1 + 10.2	Talitha; binary, 817.9 years
38	Lyn	09 18.8	+36 48	229	2.7	3.9 + 6.6	
23	UMa	09 31.5	+63 04	270	22.7	3.7 + 8.9	
54	Leo	10 55.6	+24 45	110	6.5	4.5 + 6.3	
α	UMa	11 03.7	+61 45	223	0.6	1.9 + 4.8	Dubhe; binary, 44.7 years
ξ	UMa	11 18.2	+31 32	273	1.8	4.3 + 4.8	Binary, 59.8 years
ν	UMa	11 18.5	+33 06	147	7.2	3.5 + 9.9	
57	UMa	11 29.1	+39 20	359	5.4	5.3 + 8.3	

Planetary nebula

NGC/IC	Other	Con	RA h m	Dec ° '	Mag p	Diam "	Mag*	
3587	M97	UMa	11 14.8	+55 01	12.0	194	15.9	Owl Nebula

Galaxies

NGC/IC	Other	Con	RA h m	Dec ° '	Mag	Size '	Type	
2683		Lyn	08 52.7	+33 25	9.7	9.3 × 2.5	Sb	
2681		UMa	08 53.5	+51 19	10.3	3.8 × 3.5	Sa	
2685		UMa	08 55.6	+58 44	11.1	5.2 × 3.0	Sb	
2770		Lyn	09 09.6	+33 07	12.1	3.7 × 1.3	Sc	
2768		UMa	09 11.6	+60 02	10.0	6.3 × 2.8	E5	
2776		Lyn	09 12.2	+44 57	11.6	2.9 × 2.7	Sc	
2805		UMa	09 20.3	+64 06	11.3	6.3 × 5.0	SBd	
2841		UMa	09 22.0	+50 58	9.3	8.1 × 3.8	Sb	
2903		Leo	09 32.2	+21 30	8.9	12.6 × 6.6	Sb	
3079		UMa	10 02.0	+55 41	10.7	7.6 × 1.7	Sb	
3190		Leo	10 18.1	+21 50	11.0	4.6 × 1.8	Sb	
3184		UMa	10 18.3	+41 25	9.8	6.9 × 6.8	Sc	
3193		Leo	10 18.4	+21 54	10.9	2.8 × 2.6	E0	
3198		UMa	10 19.9	+45 33	10.4	8.3 × 3.7	Sc	
3245		LMi	10 27.3	+28 30	10.8	3.2 × 1.9	E5	
3294		LMi	10 36.3	+37 20	11.7	3.3 × 1.8	Sc	
3310		UMa	10 38.7	+53 30	10.9	3.6 × 3.0	SBb	
3319		UMa	10 39.2	+41 41	11.3	6.8 × 3.9	SBc	
3344		LMi	10 43.5	+24 55	10.0	6.9 × 6.5	Sc	
3359		UMa	10 46.6	+63 13	10.5	6.8 × 4.3	SBc	
3430		LMi	10 52.2	+32 57	11.5	3.9 × 2.3	Sc	
3432		LMi	10 52.5	+36 37	11.3	6.2 × 1.5	SB	
3486		LMi	11 00.4	+28 58	10.3	6.9 × 5.4	Sc	
3504		LMi	11 03.2	+27 58	11.1	2.7 × 2.2	Sb	
3556	M108	UMa	11 11.5	+55 40	10.1	8.3 × 2.5	Sc	
—	U6253	Leo	11 13.5	+22 10	11.5	14.5 × 12.9	dE0	Leo II
3583		UMa	11 14.2	+48 19	11.7	2.8 × 2.0	Sc	
3610		UMa	11 18.4	+58 47	10.8	3.2 × 2.5	E2	
3613		UMa	11 18.6	+58 00	11.6	3.6 × 2.0	E5	
3631		UMa	11 21.0	+53 10	10.4	4.6 × 4.1	Sc	
3646		Leo	11 21.7	+20 10	11.2	3.9 × 2.6	Sc	
3642		UMa	11 22.3	+59 05	11.1	5.8 × 4.9	Sc	
3665		UMa	11 24.7	+38 46	10.8	3.2 × 2.6	E2	
3675		UMa	11 26.1	+43 35	10.9	5.9 × 3.2	Sb	
3690		UMa	11 28.5	+58 33	12.0	2.4 × 1.9	S	
3718		UMa	11 32.6	+53 04	10.5	8.7 × 4.5	SBa	
3726		UMa	11 33.3	+47 02	10.4	6.0 × 4.5	Sc	
3738		UMa	11 35.8	+54 31	11.7	2.6 × 2.0	Pec	
3769		UMa	11 37.7	+47 54	11.8	3.20 × 1.1	Sb	
3877		UMa	11 46.1	+47 30	11.6	5.4 × 1.5	Sb	
3893		UMa	11 48.6	+48 43	11.1	4.4 × 2.8	Sc	
3898		UMa	11 49.2	+56 05	10.8	4.4 × 2.6	Sb	
3941		UMa	11 52.9	+36 59	11.4	3.8 × 2.5	E3	
3945		UMa	11 53.2	+60 41	10.6	5.5 × 3.6	SBa	
3949		UMa	11 53.7	+47 52	11.0	3.0 × 1.8	Sb	
3953		UMa	11 53.8	+52 20	10.1	6.6 × 3.6	Sb	
3982		UMa	11 56.5	+55 08	11.7	2.5 × 2.2	Sb	
3992	M109	UMa	11 57.6	+53 23	9.8	7.6 × 4.9	SBb	
3998		UMa	11 57.9	+55 27	10.6	3.1 × 2.5	E2	
I750		Leo	11 58.9	+42 43	11.8	2.9 × 1.4	Sb	
4026		UMa	11 59.4	+50 58	11.7	5.1 × 1.4	S0	

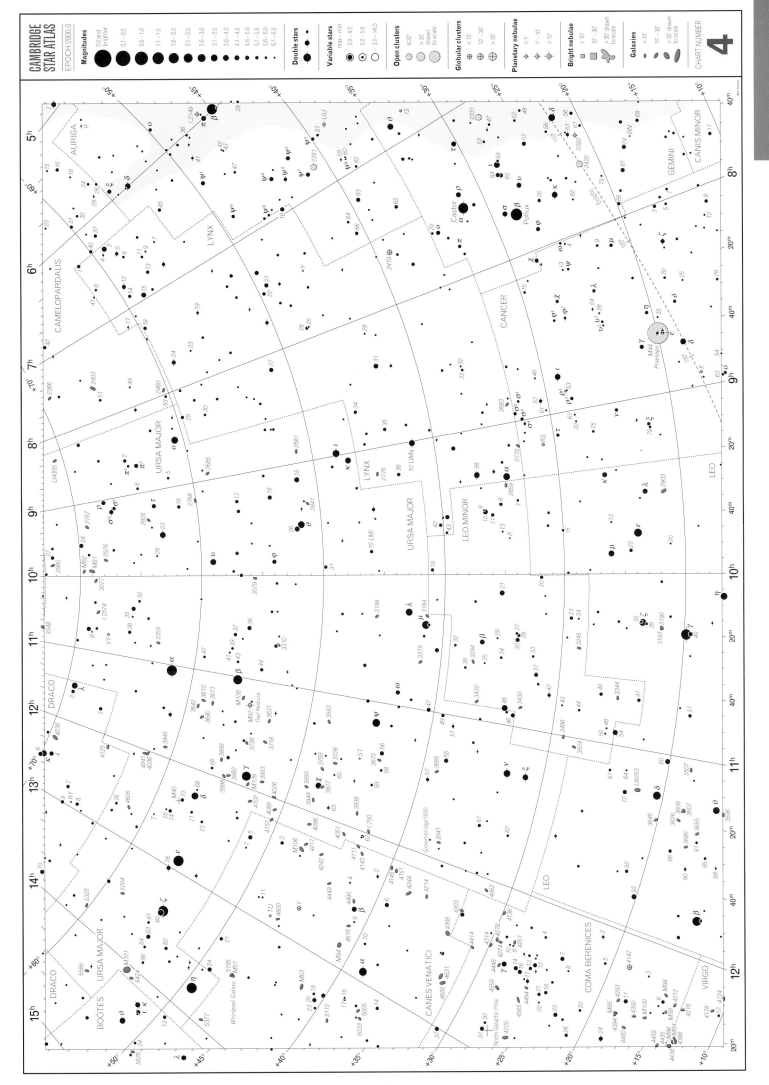

CAMBRIDGE
STAR ATLAS
EPOCH 2000.0

Magnitudes

0.0 and brighter
0.1–0.5
0.6–1.0
1.1–1.5
1.6–2.0
2.1–2.5
2.6–3.0
3.1–3.5
3.6–4.0
4.1–4.5
4.6–5.0
5.1–5.5
5.6–6.0
6.1–6.5

Double stars

Variable stars
max–min
3.0–4.2
3.0–5.9
3.0–>6.5

Open clusters
<30'
>30' drawn to scale

Globular clusters
<10'
10'–30'
>30'

Planetary nebulae
<1'
1'–10'
>10'

Bright nebulae
<30'
10'–30'
>30' drawn to scale

Galaxies
<10'
10'–30'
>30' drawn to scale

CHART NUMBER
4

Chart 5 RA 12ʰ to 16ʰ, declination +65° to +20°

Variable stars

Star	Con	RA h m	Dec ° '	Range	Type	Period (days)	Spectrum
T	UMa	12 36.4	+59 29	6.6–13.4	M	256.5	M
Y	CVn	12 45.1	+45 26	5.2–6.6	SR	257	K
TU	CVn	12 54.9	+47 12	5.6–6.6	SR	50	M
FS	Com	13 06.4	+22 37	5.3–6.1	SR	58:	M
R	CVn	13 49.0	+39 33	6.5–12.9	M	328.53	M
ZZ	Boo	13 56.2	+25 55	5.8–6.4	EW	4.99	G
V	Boo	14 29.8	+38 52	7.0–12.0	SR	258	M
R	Boo	14 37.2	+26 44	6.2–13.1	M	233.4	M
W	Boo	14 43.4	+26 32	4.7–5.4	SR	450:	M
44	Boo	15 03.8	+47 39	5.8–6.4	EW	0.27	G+G i Boo
S	CrB	15 21.4	+31 22	5.8–14.1	M	360.3	M
R	CrB	15 48.6	+28 09	5.7–14.8	RCB	—	G
T	CrB	15 59.5	+25 55	2.0–10.8	RN	29000:	M

Double stars

Star	Con	RA h m	Dec ° '	PA °	Sep "	Magnitudes	
2	CVn	12 16.1	+40 40	260	11.4	5.8 + 8.1	
α	CVn	12 56.0	+38 19	229	19.4	2.9 + 5.5	Cor Caroli
78	UMa	13 00.7	+56 22	69	1.5	5.0 + 7.4	Binary, 115.7 years
ζ	UMa	13 23.9	+54 56	152	14.4	2.3 + 4.0	Mizar
				71	708.7	4.0	80 UMa, Alcor
κ	Boo	14 13.5	+51 47	236	13.4	4.6 + 6.6	
ε	Boo	14 45.0	+27 04	339	2.8	2.5 + 4.9	Izar
39	Boo	14 49.7	+48 43	45	2.9	6.2 + 6.9	
44	Boo	15 03.8	+47 39	53	2.2	5.3 + 6.2 v	Binary, 225 years; variable
η	CrB	15 23.2	+30 17	63	0.8	5.6 + 5.9	Binary, 41.6 years
μ¹, μ²	Boo	15 24.5	+37 23	171	108.3	4.3 + 6.5 d	Alkalurops
μ²	Boo	15 24.5	+37 21	8	2.3	7.0 + 7.6	Binary, 260.1 years

Globular clusters

NGC/IC	Other	Con	RA h m	Dec ° '	Mag	Diam '
5272	M3	CVn	13 42.2	+28 23	6.4	16.2
5466		Boo	14 05.5	+28 32	9.1	11.0

Galaxies

NGC/IC	Other	Con	RA h m	Dec ° '	Mag	Size	Type	
4036		UMa	12 01.4	+61 54	10.6	4.5 × 2.0	E6	
4041		UMa	12 02.2	+62 08	11.1	2.8 × 2.7	Sc	
4051		UMa	12 03.2	+44 32	10.3	5.0 × 4.0	Sc	
4062		UMa	12 04.1	+31 54	11.2	4.3 × 2.0	Sb	
4088		UMa	12 05.6	+50 33	10.5	5.8 × 2.5	Sc	
4096		UMa	12 06.0	+47 29	10.6	6.5 × 2.0	Sc	
4102		UMa	12 06.4	+52 43	12.3	3.2 × 1.9	Sc	
4111		CVn	12 07.1	+43 04	10.8	4.8 × 1.1	S0	
4136		Com	12 09.3	+29 56	11.4	4.1 × 3.9	Sc	
4143		CVn	12 09.6	+42 32	12.1	2.9 × 1.8	E4	
4145		CVn	12 10.0	+39 53	11.0	5.8 × 4.4	Sc	
4151		CVn	12 10.5	+39 24	10.4	5.9 × 4.4	Pec	
4157		UMa	12 11.1	+50 29	11.7	6.9 × 1.7	Sb	
4214		CVn	12 15.6	+36 20	9.7	7.9 × 6.3	Irr	
4217		CVn	12 15.8	+47 06	11.9	5.5 × 1.8	Sb	
4242		CVn	12 17.5	+45 37	11.0	4.8 × 3.8	S	
4244		CVn	12 17.5	+37 49	10.2	16.2 × 2.5	S	
4251		Com	12 18.1	+28 10	11.6	4.2 × 1.9	E7	
4258	M106	CVn	12 19.0	+47 18	8.3	18.2 × 7.90	Sb	
4274		Com	12 19.8	+29 37	10.4	6.9 × 2.8	Sb	
4278		Com	12 20.1	+29 17	10.2	3.6 × 3.5	E1	
4314		Com	12 22.6	+29 53	10.5	4.8 × 4.3	SBa	
4395		CVn	12 25.8	+33 33	10.2	12.9 × 11.0	S	
4414		Com	12 26.4	+31 13	10.3	3.6 × 2.2	Sc	
4448		Com	12 28.2	+28 37	11.1	4.0 × 1.6	Sb	
4449		CVn	12 28.2	+44 06	9.4	5.1 × 3.7	Irr	
4490		CVn	12 30.6	+41 38	9.8	5.9 × 3.1	Sc	
4494		Com	12 31.4	+25 47	9.7	4.8 × 3.8	E1	
4559		Com	12 36.0	+27 58	9.9	10.5 × 4.9	Sc	
4565		Com	12 36.3	+25 59	9.6	16.2 × 2.8	SBc	
4605		UMa	12 40.0	+61 37	11.0	5.5 × 2.3	Sc	
4618		CVn	12 41.5	+41 09	10.8	4.4 × 3.8	Sc	
4631		CVn	12 42.1	+32 32	9.3	15.1 × 3.3	Sc	
4656		CVn	12 44.0	+32 10	10.4	13.8 × 3.3	Sc	
4725		Com	12 50.4	+25 30	9.2	11.0 × 7.9	SBb	
4736	M94	CVn	12 50.9	+41 07	8.2	11.0 × 9.1	Sb	
4800		CVn	12 54.6	+46 32	12.3	1.8 × 1.4	Sb	
4826	M64	Com	12 56.7	+21 41	8.5	9.3 × 5.4	Sb	Black-Eye Galaxy
5005		CVn	13 10.9	+37 03	9.8	5.4 × 2.7	Sb	
5033		CVn	13 13.4	+36 36	10.1	10.5 × 5.6	Sb	
5055	M63	CVn	13 15.8	+42 02	8.6	12.3 × 7.6	Sb	
5112		CVn	13 21.9	+38 44	11.9	3.9 × 2.9	Sc	
5204		UMa	13 29.6	+58 25	11.3	4.8 × 3.0	Sc	
5194	M51	CVn	13 29.9	+47 12	8.4	11.0 × 7.8	Irr	Whirlpool Galaxy
5195		CVn	13 30.0	+47 16	9.6	5.4 × 4.30	Pec	Companion of M51
5322		UMa	13 49.3	+60 12	10.0	5.5 × 3.9	E2	
5371		CVn	13 55.7	+40 28	10.8	4.4 × 3.6	Sb	
5377		CVn	13 56.3	+47 14	11.2	4.6 × 2.7	Sa	
5457	M101	UMa	14 03.2	+54 21	7.7	26.9 × 26.3	Sc	Pinwheel Galaxy
5474		UMa	14 05.0	+53 40	10.9	4.5 × 4.2	Sc	
5585		UMa	14 19.8	+56 44	10.9	5.5 × 3.7	S	
5676		Boo	14 32.8	+49 28	10.9	3.9 × 2.0	Sc	
5866		Dra	15 06.5	+55 46	10.0	5.2 × 2.3	E6	
5879		Dra	15 09.8	+57 00	11.5	4.4 × 1.7	Sb	
5907		Dra	15 15.9	+56 19	10.4	12.3 × 1.8	Sc	
5982		Dra	15 38.7	+59 21	11.1	2.9 × 2.2	E3	
5985		Dra	15 39.6	+59 20	11.0	5.5 × 3.2	Sb	
6015		Dra	15 51.4	+62 19	11.2	5.4 × 2.3	Sc	

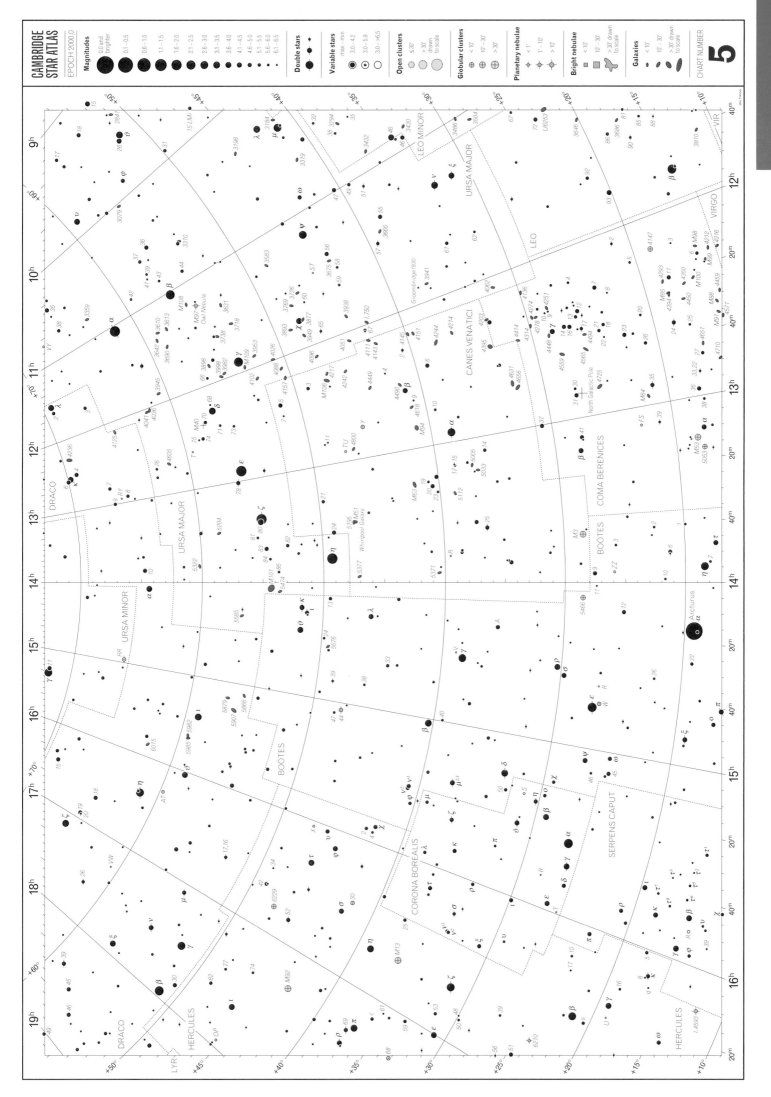

Chart 6   RA 16ʰ to 20ʰ, declination +65° to +20°

Variable stars

Star	Con	RA h m	Dec ° '	Range	Type	Period (days)	Spectrum	
X	Her	16 02.7	+47 14	6.3–7.4	SR	95.0	M	
AT	Dra	16 17.3	+59 45	5.3–6.0	Irr	—	M	
30	Her	16 28.6	+41 53	4.3–6.3	SR	70:	M	g Her
VW	Dra	17 16.5	+60 40	6.0–6.5	SR	170:	K	
68	Her	17 17.3	+33 06	4.7–5.4	EB	2.05	B	u Her
OP	Her	17 56.8	+45 21	5.9–6.7	Irr	—	M	
XY	Lyr	18 38.1	+39 40	5.8–6.4	Irr	—	M	
β	Lyr	18 50.1	+33 22	3.3–4.3	EB	12.94	B+A	Sheliak
R	Lyr	18 55.3	+43 57	3.9–5.0	SR	46.0	M	
AF	Cyg	19 30.2	+46 09	6.4–8.4	SR	94.1	M	
U	Vul	19 36.6	+20 20	6.8–7.5	Cep	7.99	F–G	
R	Cyg	19 36.8	+50 12	6.1–14.2	M	426.4	S	
V1143	Cyg	19 38.7	+54 58	5.9–6.4	EA	7.64	F	
RT	Cyg	19 43.6	+48 47	6.4–12.7	M	190.3	M	
V973	Cyg	19 44.8	+40 43	6.1–6.6	SR	40:	M	
SU	Cyg	19 44.8	+29 16	6.5–7.2	Cep	3.85	F	
χ	Cyg	19 50.6	+32 55	3.3–14.2	M	406.9	S	
V449	Cyg	19 53.3	+33 57	7.4–9.0 p	Irr	—	M	

Open clusters

NGC/IC	Other	Con	RA h m	Dec ° '	Mag	Diam '	N*	
—	Steph 1	Lyr	18 53.5	+36 55	3.8	20	15	δ Lyrae Cluster
—	Cr 399	Vul	19 25.4	+20 11	3.6	60	40	Brocchi's Cluster
6811		Cyg	19 38.2	+46 34	6.8	13	70	
6819		Cyg	19 41.3	+40 11	7.3	5		
6823		Vul	19 43.1	+23 18	7.1	12	30	In nebula 6820
6830		Vul	19 51.0	+23 04	7.9	12	20	
6834		Cyg	19 52.2	+29 25	7.8	5	50	

Globular clusters

NGC/IC	Other	Con	RA h m	Dec ° '	Mag	Diam '	
6205	M13	Her	16 41.7	+36 28	5.9	16.6	Hercules Cluster
6229		Her	16 47.0	+47 32	9.4	4.5	
6249	M92	Her	17 17.1	+43 08	6.5	11.2	
6779	M56	Lyr	19 16.6	+30 11	8.2	7.1	

Planetary nebulae

NGC/IC	Other	Con	RA h m	Dec ° '	Mag p	Diam "	Mag*	
6210		Her	16 44.5	+23 49	9.3	14	12.9	
—	PK51+9.1	Her	18 49.7	+20 51	12.2	3	13.0	
6720	M57	Lyr	18 53.6	+33 02	9.7	70/150	14.8	Ring Nebula
—	PK64+5.1	Cyg	19 34.8	+30 31	9.6	8	10.0	
6826		Cyg	19 44.8	+50 31	9.8	30/140	10.4	Blinking Planetary
6853	M27	Vul	19 59.6	+22 43	7.6	350/910	13.9	Dumbbell Nebula

Double stars

Star	Con	RA h m	Dec ° '	PA °	Sep "	Magnitudes	
η	Dra	16 24.0	+61 31	142	5.2	2.7+8.7	
17	Dra	16 36.2	+52 55	108	3.4	5.4+6.4	16 Dra
				194	90.3	5.5	
ζ	Her	16 41.3	+31 36	12	0.8	2.9+5.5	Binary, 34.5 years
μ	Dra	17 05.3	+54 28	8	1.9	5.7+5.7	Binary, 482 years
ρ	Dra	17 23.7	+37 09	316	4.1	4.6+5.6	
ν	Dra	17 32.2	+55 11	312	61.9	4.9+4.9	
μ	Her	17 46.5	+27 43	247	33.8	3.4+10.1	
90	Her	17 53.3	+40 00	116	1.6	5.2+8.5	
95	Her	18 01.5	+21 36	258	6.3	5.0+5.1	
100	Her	18 07.8	+26 06	183	14.2	5.9+6.0	
39	Dra	18 23.9	+58 48	351	3.1	5.0+8.0	
				21	88.9	7.4	5 more fainter components
ε	Lyr	18 44.3	+39 40	207.7	173	4.7+4.6	ε¹, ε²
				350	2.6	5.0+6.1	ε¹, binary, 1165 years
				82	2.3	5.2+5.5	ε², binary, 585 years
ζ	Lyr	18 44.8	+37 36	150	43.7	4.3+5.9	ζ¹, ζ²
β	Lyr	18 50.1	+33 22	149	45.7	3.4v+8.6	Sheliak; variable
o	Dra	18 51.2	+59 23	326	34.2	4.8+7.8	
η	Lyr	19 13.8	+39 09	82	28.1	4.4+9.1	
2	Vul	19 17.7	+23 02	127	1.8	5.4+9.2	ES Vul
α, 8	Vul	19 28.7	+24 40	28	413.7	4.4+5.8	
β	Cyg	19 30.7	+27 58	54	34.4	3.1+5.1	Albireo
δ	Cyg	19 45.0	+45 08	221	2.5	2.9+6.3	Binary, 827.6 years
17	Cyg	19 46.4	+33 44	69	26.0	5.0+9.2	
ψ	Cyg	19 55.6	+52 26	178	3.2	4.9+7.4	

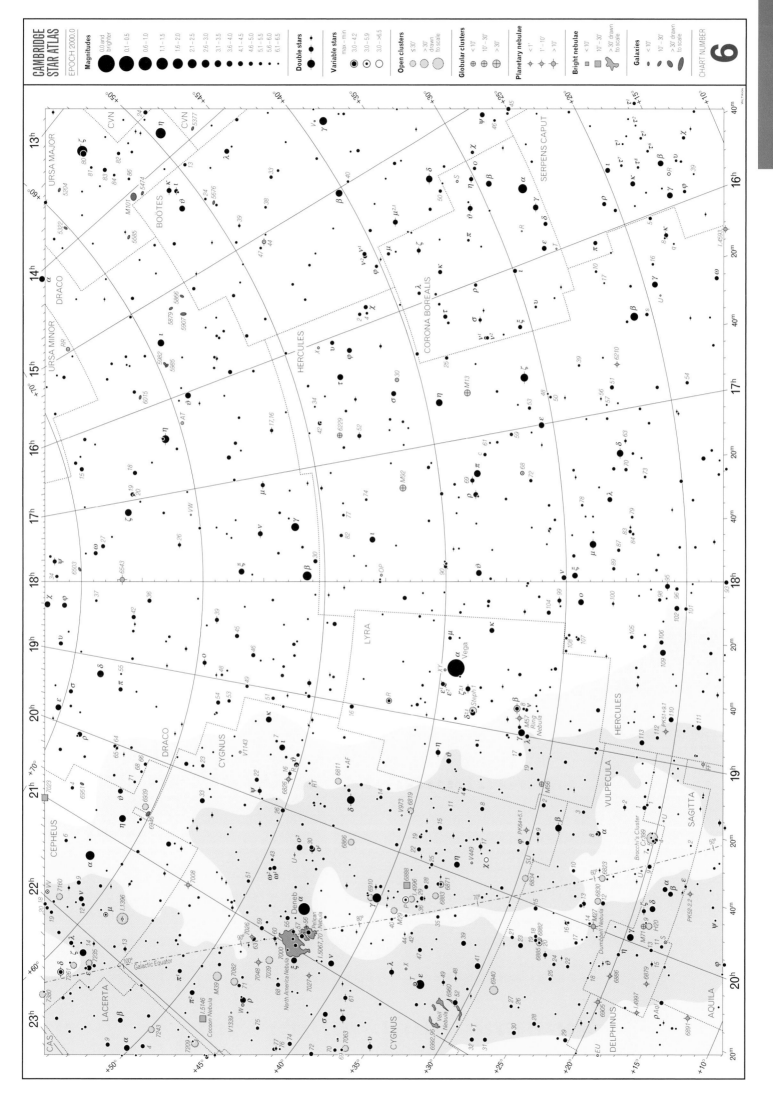

Chart 7 RA 20ʰ to 0ʰ, declination +65° to +20°

Variable stars

Star	Con	RA h m	Dec ° '	Range	Type	Period (days)	Spectrum
P	Cyg	20 17.8	+38 02	3.0–6.0	SD	—	B
U	Cyg	20 19.6	+47 54	5.9–12.1	M	462.4	M
X	Cyg	20 43.4	+35 35	5.9–6.7	Cep	16.39	F–G
T	Cyg	20 47.2	+34 22	5.0–5.5	Irr	—	K
T	Vul	20 51.5	+28 15	5.4–6.1	Cep	4.43	F–G
W	Cyg	21 36.0	+45 22	5.0–7.6	SR	126.3	M
V460	Cyg	21 42.0	+35 31	5.6–7.0	Irr	—	M
V1339	Cyg	21 42.1	+45 46	5.9–7.1	SR	35:	M
V	Cep	21 43.5	+58 47	3.4–5.1	SR	730	M
VV	Cep	21 56.7	+63 38	4.8–5.4	EA	7430	M
AR	Lac	22 08.7	+45 45	6.1–6.8	EA	1.98	G+K
δ	Cep	22 29.2	+58 25	3.9–4.4	Cep	5.37	F–G
V509	Cas	23 00.1	+56 57	4.8–5.5	SR	—	F–K
β	Peg	23 03.8	+28 05	2.3–2.7	Irr	—	M Scheat
Z	And	23 33.7	+48 49	8.0–12.4 p	ZA	—	M
ρ	Cas	23 54.4	+57 30	4.1–6.2	SR	320	F–K
R	Cas	23 58.4	+51 24	4.7–13.5	M	430.5	M

Double stars

Star	Con	RA h m	Dec ° '	PA °	Sep "	Magnitudes	
52	Cyg	20 45.7	+30 43	67	6.0	4.2+6.4	In Veil Nebula 6920
61	Cyg	21 06.9	+38 45	150	30.3	5.2+6.0	Binary, 653.3 years; large p.m.
τ	Cyg	21 14.8	+38 03	306	0.8	3.8+6.4	Binary, 49.9 years
υ	Cyg	21 17.9	+34 54	220	15.1	4.4+10.0	
μ	Cyg	21 44.1	+28 45	320	1.2	4.8+6.1	Binary, 507.5 years
ξ	Cep	22 03.8	+64 38	274	8.2	4.4+6.5	Binary, 3800 years
δ	Cep	22 29.2	+58 25	191	41.0	3.9 v+7.5	Variable
8	Lac	22 35.9	+39 38	186	22.4	5.7+6.5	
				169	48.8	10.5	
				144	81.8	9.3	
				239	336.6	7.8	
72	Peg	23 34.0	+31 20	97	0.5	5.7+5.8	Binary, 241.2 years
78	Peg	23 44.0	+29 22	235	1.0	5.0+8.1	
6	Cas	23 48.8	+62 13	193	1.6	5.5+8.0	
σ	Cas	23 59.0	+55 45	326	3.0	5.0+7.1	

Open clusters

NGC/IC	Other	Con	RA h m	Dec ° '	Mag	Diam '	N*	
6866		Cyg	20 03.7	+44 00	7.6	7	80	
6871		Cyg	20 05.9	+35 47	5.2	20	15	
6883		Cyg	20 11.3	+35 51	8.0	15	30	
6882		Vul	20 11.7	+26 33	8.1	18		Near cluster 6885
6885		Vul	20 12.0	+26 29	5.7	7	30	Contains 20 Vul
I.4996		Cyg	20 16.5	+37 38	7.3	6	15	
6910		Cyg	20 23.1	+40 47	7.4	8	50	
6913	M29	Cyg	20 23.9	+38 32	6.6	7	50	
6939		Cep	20 31.4	+60 38	7.8	8	80	
6940		Vul	20 34.6	+28 18	6.3	31	60	
7039		Cyg	21 11.2	+45 39	7.6	25	50	
7063		Cyg	21 24.4	+36 30	7.0	8	12	
7082		Cyg	21 29.4	+47 05	7.2	25		
7092	M39	Cyg	21 32.2	+48 26	4.6	32	30	
I.1396		Cep	21 39.1	+57 30	3.5	50	50	
7160		Cep	21 53.7	+62 36	6.1	7	12	
7209		Lac	22 05.2	+46 30	6.7	25	25	
7235		Cep	22 12.6	+57 17	7.7	4	30	
7243		Lac	22 15.3	+49 53	6.4	21	40	
7261		Cep	22 20.4	+58 05	8.4	6	30	
7380		Cep	22 47.0	+58 06	7.2	12	40	
7510		Cep	23 11.5	+60 34	7.9	4	60	
7654	M52	Cas	23 24.2	+61 35	6.9	13	100	
7686		And	23 30.2	+49 08	5.6	15	20	
7789		Cas	23 57.0	+56 44	6.7	16	300	
7790		Cas	23 58.4	+61 13	8.5	17	40	

Bright diffuse nebulae

NGC/IC	Other	Con	RA h m	Dec ° '	Type	Diam '	Mag*	
6888		Cyg	20 12.0	+38 21	E	20×10	7.4	
6960		Cyg	20 45.7	+30 43	E	70×6		Veil Nebula SNR
		Cyg	20 48.5	+31 09	E	45×30		Veil Nebula SNR
I.5067, I.5070		Cyg	20 50.8	+44 21	E	80×70		Pelican Nebula
6992		Cyg	20 56.4	+31 43	E	60×8		Veil Nebula SNR
6995		Cyg	20 57.1	+31 13	E	12		Veil Nebula SNR
7000		Cyg	20 58.8	+44 20	E	120×100	6.0	North America Nebula
I.5146		Cyg	21 53.5	+47 16	E	12×12	10.0	Cocoon Nebula
7635		Cas	23 20.7	+61 12	E	15×8	6.9	Bubble Nebula

Planetary nebulae

NGC/IC	Other	Con	RA h m	Dec ° '	Mag p	Diam "	
6905		Del	20 22.4	+20 07	11.9	46/100	
7008		Cyg	21 00.6	+54 33	13.3	83	
7026		Cyg	21 06.3	+47 51	12.7	21	
7027		Cyg	21 07.1	+42 14	10.4	15	
7048		Cyg	21 14.2	+46 16	11.3	61	
7662		And	23 25.9	+42 33	9.2	20/130	

Galaxies

NGC/IC	Other	Con	RA h m	Dec ° '	Mag	Size '	Type
6946		Cep	20 34.8	+60 09	8.9	11.0×9.8	Sc
7217		Peg	22 07.9	+31 22	10.2	3.7×3.2	Sb
7331		Peg	22 37.1	+34 25	9.5	10.7×4.0	Sb
7332		Peg	22 37.4	+23 48	11.0	4.2×1.3	E7
7457		Peg	23 01.0	+30 09	10.8	4.4×2.5	E
7640		And	23 22.1	+40 51	10.9	10.7×2.5	SBb
7741		Peg	23 43.9	+26 05	11.4	4.0×2.8	SBc

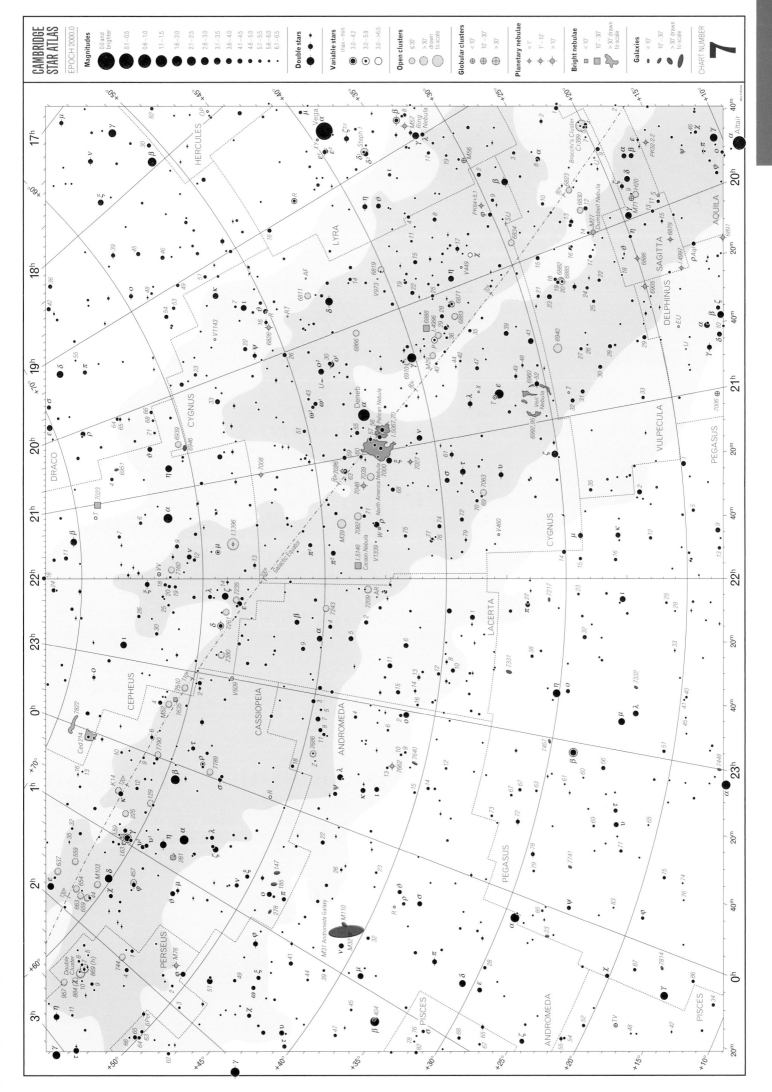

Chart 8 RA 0ʰ to 4ʰ, declination +20° to –20°

Variable stars

Star	Con	RA h m	Dec ° '	Range	Type	Period (days)	Spectrum	
TV	Psc	00 28.0	+17 54	4.7–5.4	SR	70	M	
o	Cet	02 19.3	–02 59	2.0–10.1	M	332.0	M	Mira
U	Cet	02 33.7	–13 09	6.8–13.4	M	234.8	M	
Z	Eri	02 47.9	–12 28	5.6–7.2	SR	80	M	
RR	Eri	02 52.2	–08 16	6.3–8.1	SR	97	M	

Double stars

Star	Con	RA h m	Dec ° '	PA °	Sep "	Magnitudes	Spectrum
51	Psc	00 32.4	+06 57	83	27.5	5.7 + 9.5	
26	Cet	01 03.8	+01 22	253	16.0	6.2 + 8.6	
ζ	Psc	01 13.7	+07 35	63	23.0	5.6 + 6.5	
37	Cet	01 14.4	–07 55	331	49.7	5.2 + 8.7	
χ	Cet	01 49.6	–10 41	250	183.8	4.9 + 6.9	
γ	Ari	01 53.5	+19 18	0	7.8	4.8 + 4.8	Mesartim
α	Psc	02 02.0	+02 46	272	1.8	4.2 + 5.1	Binary, 933.1 years
66	Cet	02 12.8	–02 24	234	16.5	5.7 + 7.5	
84	Cet	02 41.2	–02 42	310	4.0	5.8 + 9.0	
γ	Ari	02 43.3	+03 14	294	2.8	3.5 + 7.3	
π	Ari	02 49.3	+17 28	120	3.2	5.2 + 8.7	
ρ³	Eri	03 02.7	–07 41	75	1.8	5.3 + 9.5	
95	Cet	03 18.4	–00 56	250	1.1	5.6 + 7.5	Binary, 217.2 years

Planetary nebula

NGC/IC	Other	Con	RA h m	Dec ° '	Mag p	Diam "	Mag*
246		Cet	00 47.0	–11 53	8.0	225	11.9

Galaxies

NGC/IC	Other	Con	RA h m	Dec ° '	Mag	Size '	Type
7814		Peg	00 03.3	+16 09	10.5	6.3×2.6	Sb
128		Psc	00 29.3	+02 52	11.6	3.4×1.0	S0
157		Cet	00 34.8	–08 24	10.4	4.3×2.90	Sc
210		Cet	00 40.6	–13 52	10.9	5.4×3.7	Sb
255		Cet	00 47.8	–11 28	11.8	3.1×2.8	Sb
309		Cet	00 56.7	–09 55	11.8	3.1×2.7	Sc
337		Cet	00 59.8	–07 35	11.6	2.8×2.0	Sc
1.1613		Cet	01 04.8	+02 07	9.3	12.0×11.2	Irr
428		Cet	01 12.9	+00 59	11.4	4.1×3.2	Sc
474		Cet	01 20.1	+03 25	11.1	7.9×7.20	S0
488		Psc	01 21.8	+05 15	10.3	5.2×4.1	Sb
524		Cet	01 24.8	+09 32	10.6	3.2×3.2	E1
584		Cet	01 31.1	–06 52	10.4	3.8×2.4	E4
628	M74	Psc	01 36.7	+15 47	9.2	10.2×9.5	Sc
676		Psc	01 49.0	+05 54	11.0	4.3×1.5	Sa
720		Cet	01 53.0	–13 44	10.2	4.4×2.8	E3
772		Ari	01 59.3	+19 01	10.3	7.1×4.5	Sb
779		Cet	01 57.7	–05 58	11.1	4.1×4.5	Sb
864		Cet	02 15.5	+06 00	11.0	4.6×3.50	Sc
895		Cet	02 21.6	–05 31	11.8	3.6×2.8	Sb
936		Cet	02 27.6	–01 09	10.1	5.2×4.4	SBa
988		Cet	02 35.4	–09 21	11:	4.7×2.7	SBc
1042		Cet	02 40.4	–08 26	10.9	4.7×3.9	Sc
1052		Cet	02 41.1	–08 15	10.6	2.9×2.0	E2
1055		Cet	02 41.8	+00 26	10.6	7.6×3.0	Sb
1068	M77	Cet	02 42.7	–00 01	8.8	6.9×5.9	Sb
1073		Cet	02 43.7	+01 23	11.0	4.9×4.6	SBc
1084		Eri	02 46.0	–07 35	10.6	2.9×1.5	Sc
1087		Cet	03 46.4	–00 30	11.1	3.5×2.3	Sc
1179		Eri	03 02.6	–18 54	11.8	4.6×3.9	S
1300		Eri	03 19.7	–19 25	10.4	6.5×4.3	SBb
1337		Eri	03 28.1	–08 23	11.7	6.8×2.0	S
1407		Eri	03 40.2	–18 35	9.8	2.5×2.5	E0

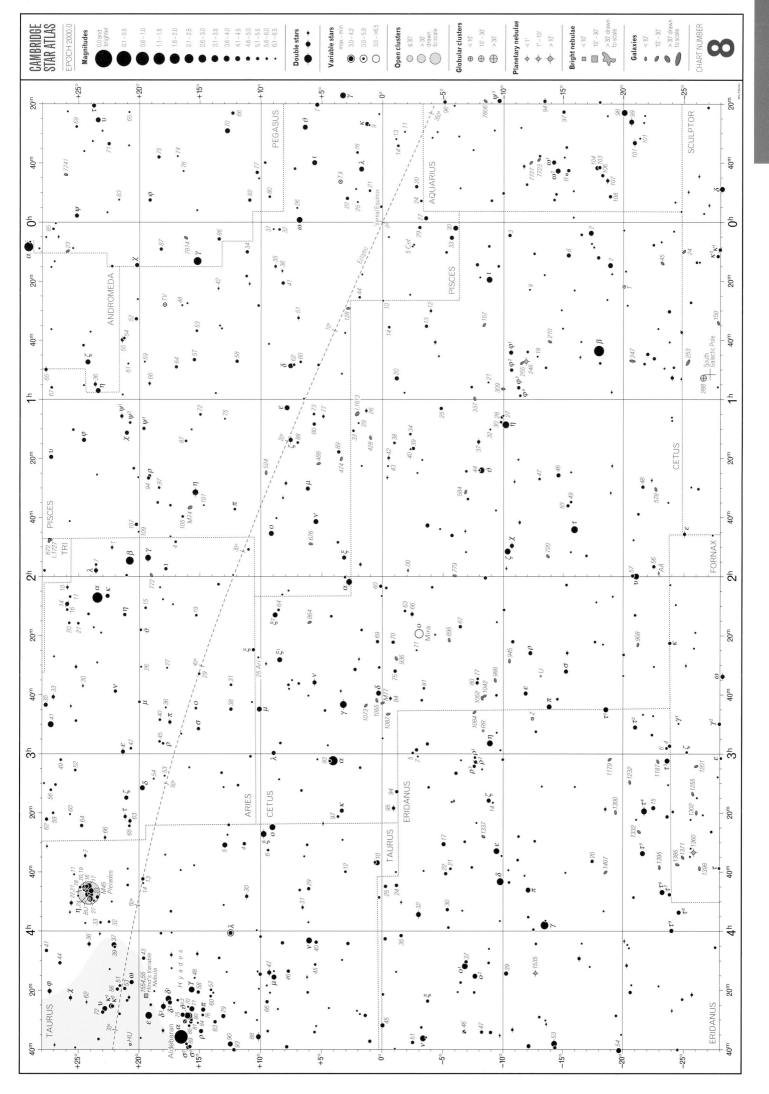

Chart 9 RA 4ʰ to 8ʰ, declination +20° to −20°

Variable stars

Star	Con	RA h m	Dec ° '	Range	Type	Period (days)	Spectrum	
λ	Tau	04 00.7	+12 29	3.3–3.8	EA	3.95	B + A	
R	Lep	04 59.6	−14 18	5.5–11.7	M	432.1	M	
W	Ori	05 05.4	+01 11	5.9–7.7	SR	212	B	
RX	Lep	05 11.4	−11 51	5.0–7.0	Irr	—	B	
CK	Ori	05 30.3	+04 12	5.9–7.1	SR	120:	M	
α	Ori	05 55.2	+07 24	0.4–1.3	SR	2110	M	Betelgeuse
V	Mon	06 22.7	−02 12	6.0–13.7	M	333.8	M	
T	Mon	06 25.2	+07 05	5.6–6.6	Cep	27.02	F – K	
BL	Ori	06 25.5	+14 43	6.3–6.90	Irr	—	M	
R	CMa	07 19.5	−16 24	5.7–6.3	EA	1.14	F	
U	Mon	07 30.8	−09 47	5.9–7.8	RV	92.26	F – K	
BN	Gem	07 37.1	+16 54	6.0–6.6 p	Irr	—	O	

Double stars

Star	Con	RA h m	Dec ° '	PA °	Sep "	Magnitudes	
47	Tau	04 13.9	+09 16	351	1.1	4.9 + 7.4	
θ¹, θ²	Tau	04 28.7	+15 52	346	337.4	3.4 + 3.8	
σ¹, σ²	Tau	04 39.3	+15 55	193	431.2	4.7 + 5.1	
55	Eri	04 43.6	−08 48	317	9.2	6.7 + 6.8	
14	Ori	05 07.9	+08 30	322	0.8	5.8 + 6.5	Binary, 198.9 years
ρ	Ori	05 13.3	+02 52	64	7.0	4.5 + 8.3	
κ	Lep	05 13.2	−12 56	358	2.6	4.5 + 7.4	
β	Ori	05 14.5	−08 12	202	9.5	0.2 + 6.8	Rigel
η	Ori	05 24.5	−02 24	80	1.5	3.8 + 4.5	
θ¹	Ori	05 35.3	−05 23	31	8.8	6.7 + 7.9	Trapezium; in Orion Nebula
				132	12.8	5.1	
				96	21.5	6.7	
θ²	Ori	05 35.4	−05 25	92	52.5	5.2 + 6.5	In Orion Nebula
ι	Ori	05 35.4	−05 55	141	11.3	2.8 + 6.9	In Orion Nebula
ζ	Ori	05 40.8	−01 57	165	2.3	1.9 + 4.0	Alnitak; binary, 1509 years
75	Ori	06 17.1	+09 57	258	62.7	5.4 + 9.5	
				159	117.3	8.5	
8	Mon	06 23.8	+04 36	27	13.4	4.5 + 6.5	= ε Mon
β	Mon	06 28.8	−07 02	132	7.3	4.7 + 5.2	
				124	10.0	6.1	
ν¹	CMa	06 36.4	−18 40	262	17.5	5.8 + 8.5	
15	Mon	06 41.0	+09 54	213	2.8	4.7 + 7.5	S Mon
				13	16.6	9.8	
				308	41.3	9.6	In cluster 2264
				139	73.9	9.9	
				222	156.0	7.7	
α	CMa	06 45.2	−16 43	150	4.6	−1.4 + 8.5	Sirius; binary 50.1 years
38	Gem	06 54.6	+13 11	145	7.1	4.7 + 7.7	Binary, 3,190 years
μ	CMa	06 56.1	−14 03	340	3.0	5.3 + 8.6	

Open clusters

NGC/IC	Other	Con	RA h m	Dec ° '	Mag	Diam	N*	
1647		Tau	04 46.0	+19 04	6.4	45	200	
1662		Ori	04 48.5	+10 56	6.4	20	35	
1807		Tau	05 10.7	+16 32	7.0	17	20	
1817		Tau	05 12.1	+16 42	7.7	16	60	
1981		Ori	05 35.2	−04 26	4.6	25	20	
2169		Ori	06 08.4	+13 57	5.9	7	30	
2194		Ori	06 13.8	+12 48	8.5	10	80	
2204		CMa	06 15.7	−18 39	8.6	13	80	
2215		Mon	06 21.0	−07 17	8.4	11	40	
2232		Mon	06 26.6	−04 45	3.9	30	20	Contains 10 Mon
2244		Mon	06 32.4	+04 52	4.80	24	100	In Rosette Nebula
2252		Mon	06 35.0	+05 23	7.7	20	30	
2286		Mon	06 47.6	−03 10	7.5	15	50	Asterism?
2301		Mon	06 51.8	+00 28	6.0	12	80	
2323	M50	Mon	07 03.20	−08 20	5.9	16	80	
2335		Mon	07 06.6	−10 05	7.2	12	35	
2343		Mon	07 08.3	−10 39	6.7	7	20	
2345		CMa	07 08.3	−13 10	7.7	12	70	
2353		Mon	07 14.6	−10 18	7.1	20	30	
2360		CMa	07 17.8	−15 37	7.2	13	80	
2374		CMa	07 24.0	−13 16	8.0	19	80	
2395		Gem	07 27.1	+13 35	8.0	12	25	
2414		Pup	07 33.3	−15 27	7.9	4	35	
2422	M47	Pup	07 36.6	−14 30	4.4	30	30	
2423		Pup	07 37.1	−13 52	6.7	19	40	
—	Mel71	Pup	07 37.5	−12 04	7.1	9	80	
2437	M46	Pup	07 41.8	−14 49	6.1	27	100	Contains pl. neb. 2438

Bright diffuse nebulae

NGC/IC	Other	Con	RA h m	Dec ° '	Type	Diam	Mag*	
1554, 1555		Tau	04 21.8	+19 32	R	Var	9.4	Hind's Variable Nebula
1973		Ori	05 35.1	−04 44	E + R	5×5	7.4	
1975		Ori	05 35.4	−04 41	E + R	10×5	10.9	
1976	M42	Ori	05 35.4	−05 27	E + R	66×60	2.9	Great Orion Nebula
1977		Ori	05 35.5	−04 52	E + R	20×10	4.6	
1982	M43	Ori	05 35.6	−05 16	E + R	20×15	6.9	Extension of M42
I.434		Ori	05 41.0	−02 24	E	60×10	2.1	Contains Horse Head Nebula (B33)
2024		Ori	05 40.7	−02 27	E	30×30	2.1	ζ Ori
2068	M78	Ori	05 46.7	+00 03	R	8×6	10.5	
2237–2239		Mon	06 32.3	+05 03	E	80×60	10.0	Rosetta Nebula
2261	R Mon	Mon	06 39.2	+08 44	E + R	2×1	10.0	Hubble's Variable Nebula
2264	S Mon	Mon	06 40.9	+09 54	E	60×30	4.5	Includes Cone Nebula
I.2177		Mon	07 05.1	−10 42	E	120×40	6.2	

Planetary nebulae

NGC/IC	Other	Con	RA h m	Dec ° '	Mag p	Diam "	Mag*	
1535		Eri	04 14.2	−12 44	9.6	18/44	12.2	
I.418		Lep	05 27.5	−12 42	10.7	12	10.6	
2438		Pup	07 41.8	−14 44	10.1	66	17.7	In open cluster 2437
2440		Pup	07 41.9	−18 13	10.8	14/32	14.3	

Galaxies

NGC/IC	Other	Con	RA h m	Dec ° '	Mag	Size '	Type
1637		Eri	04 41.5	−02 51	10.9	3.3×2.9	Sc
1723		Eri	04 59.4	−10 59	12:	3.7×2.3	SB

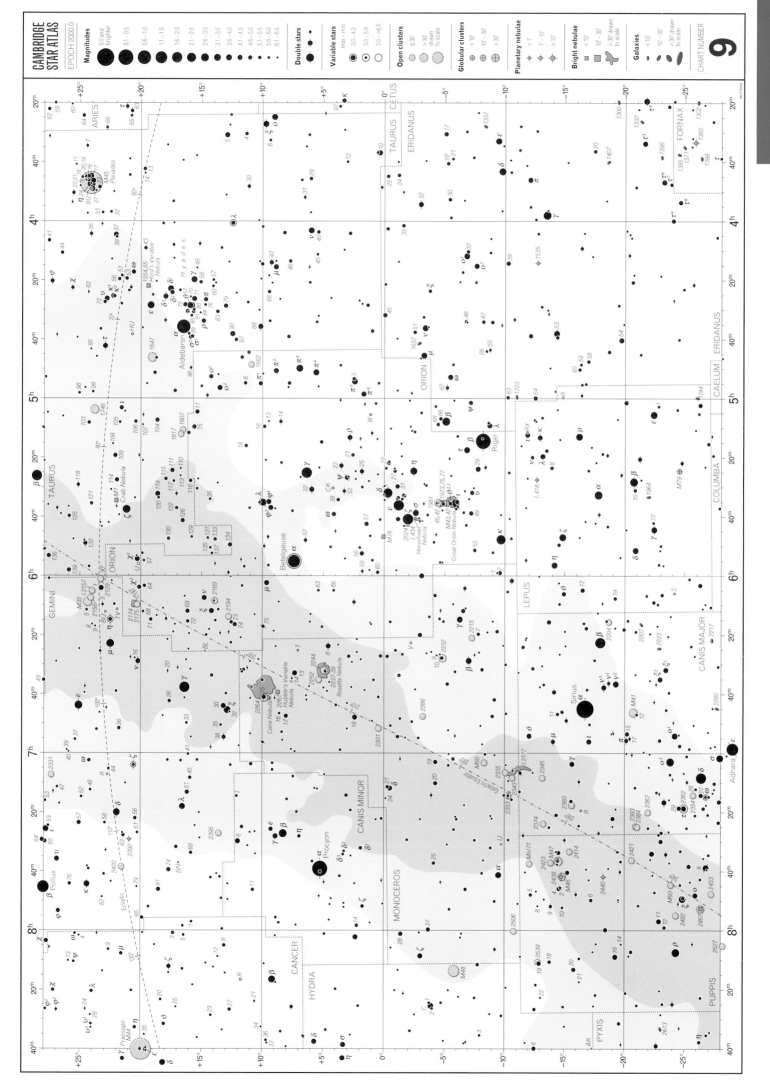

Chart 10 RA 8ʰ to 12ʰ, declination +20° to –20°

Variable stars

Star	Con	RA h m	Dec ° '	Range	Type	Period (days)	Spectrum
R	Cnc	08 16.6	+11 44	6.1–11.8	M	361.6	M
AK	Hya	08 39.9	–17 18	6.3–6.9	SR	112:	M
X	Cnc	08 55.4	+17 14	5.6–7.5	SR	195:	K
R	Leo	09 47.6	+11 26	4.4–11.3	M	312.4	M
U	Hya	10 37.6	–13 23	4.3–6.5	SR	450:	B

Double stars

Star	Con	RA h m	Dec ° '	PA °	Sep "	Magnitudes	Spectrum
ζ	Cnc	08 12.2	+17 39	72	6.0	5.6 + 6.2	Binary, 1150 years
				86	0.8	6.0	Binary, 59.7 years
ε	Hya	08 46.8	+06 25	302	2.7	3.8 + 4.7	Binary, 890 years
27	Hya	09 20.5	–09 33	211	229.4	5.0 + 6.9	
ω	Leo	09 28.5	+09 03	84	0.6	5.9 + 6.5	Binary, 118.2 years
6	Leo	09 32.0	+09 43	75	37.4	5.2 + 8.2	
γ	Sex	09 52.5	–08 06	56	0.6	5.6 + 6.1	Binary, 75.6 years
α	Leo	10 08.4	+11 58	307	176.9	1.4 + 7.7	Regulus
γ	Leo	10 20.0	+19 51	125	4.4	2.2 + 3.5	Algieba; binary, 618.6 years
				291	259.9	9.2	
				302	333.0	9.6	
49	Leo	10 35.0	+08 39	157	2.4	5.8 + 8.5	TX Leo
1	Leo	11 23.9	+10 32	117	1.7	4.0 + 6.7	Binary, 192 years
γ	Crt	11 24.9	–17 41	96	5.2	4.1 + 9.6	
90	Leo	11 34.7	+16 48	209	3.3	6.0 + 7.3	
				234	63.1	8.7	

Open clusters

NGC/IC	Other	Con	RA h m	Dec ° '	Mag	Diam '	N*	
2506		Mon	08 00.2	–10 47	7.6	7	150	
2539		Pup	08 10.7	–12 50	6.5	22	50	
2548	M48	Hya	08 13.8	–05 48	5.8	54	80	
2632	M44	Cnc	08 40.1	+19 59	3.1	95	50	Praesepe; Beehive
2682	M67	Cnc	08 50.4	+11 49	6.9	30	200	

Planetary nebula

NGC/IC	Other	Con	RA h m	Dec ° '	Diam "	Mag p	Mag*	
3242		Hya	10 24.8	–18 38	16/1250	8.6	12.0	Ghost of Jupiter

Galaxies

NGC/IC	Other	Con	RA h m	Dec ° '	Mag	Size '	Type	
2775		Cnc	09 10.3	+07 02	10.3	4.5×3.5	Sa	
2967		Sex	09 42.1	+00 20	11.6	3.0×2.9	Sc	
2974		Sex	09 42.6	–03 42	10.8	3.4×2.1	Sa	
—	U5373	Sex	10 00.0	+05 20	11.4	4.6×3.3	Irr	
3115		Sex	10 05.2	–07 43	9.2	8.3×3.2	E6	
—	U5470	Leo	10 08.4	+12 18	9.8	10.7×8.3	dE3	Leo I
3166		Sex	10 13.8	+03 26	10.6	5.2×2.7	SBa	
3169		Sex	10 14.2	+03 28	10.5	4.8×3.2	Sb	
3351	M95	Leo	10 44.0	+11 42	9.7	7.4×5.1	SBb	
3367		Leo	10 46.6	+13 45	11.5	2.3×2.1	Sc	
3368	M96	Leo	10 46.8	+11 49	9.2	7.1×5.1	Sb	
3377		Leo	10 47.7	+13 59	10.2	4.4×2.7	E5	
3379	M105	Leo	10 47.8	+12 35	9.3	4.5×4.0	E1	
3384		Leo	10 48.3	+12 38	10.0	5.9×2.6	E7	
3412		Leo	10 50.9	+13 25	10.6	3.6×2.0	E5	
3489		Leo	11 00.3	+13 54	10.3	3.7×2.1	E6	
3507		Leo	11 03.5	+18 08	11.4	3.5×3.0	SBb	
3521		Leo	11 05.8	–00 02	8.9	9.5×5.0	Sb	
3593		Leo	11 14.6	+12 49	11.0	5.8×2.5	Sb	
3596		Leo	11 15.1	+14 47	11.6	4.2×4.1	Sc	
3607		Leo	11 16.9	+18 03	10.0	3.7×3.2	E1	
3623	M65	Leo	11 18.9	+13 05	9.3	10.0×3.3	Sb	
3626		Leo	11 20.1	+18 21	10.9	3.1×2.2	Sb	
3627	M66	Leo	11 20.2	+12 59	9.0	8.7×4.4	Sb	
3628		Leo	11 20.3	+13 36	9.5	14.8×3.6	Sb	
3640		Leo	11 21.1	+03 14	10.3	4.1×3.4	E1	
3655		Leo	11 22.9	+16 35	11.6	1.6×1.1	S:	
3672		Crt	11 25.0	–09 48	11.5	4.1×2.1	Sb	
3686		Leo	11 27.7	+17 13	11.4	3.3×2.6	Sc	
3810		Leo	11 41.0	+11 28	10.8	4.3×3.1	Sc	
3818		Vir	11 42.0	–06 09	11.8	2.1×1.4	E2	
3887		Crt	11 47.1	–16 51	11.0	3.3×2.7	Sc	
3962		Crt	11 54.7	–13 58	10.6	2.9×2.6	E2	
4027		Crv	11 59.5	–19 16	11.1	3.0×2.3	Sc	

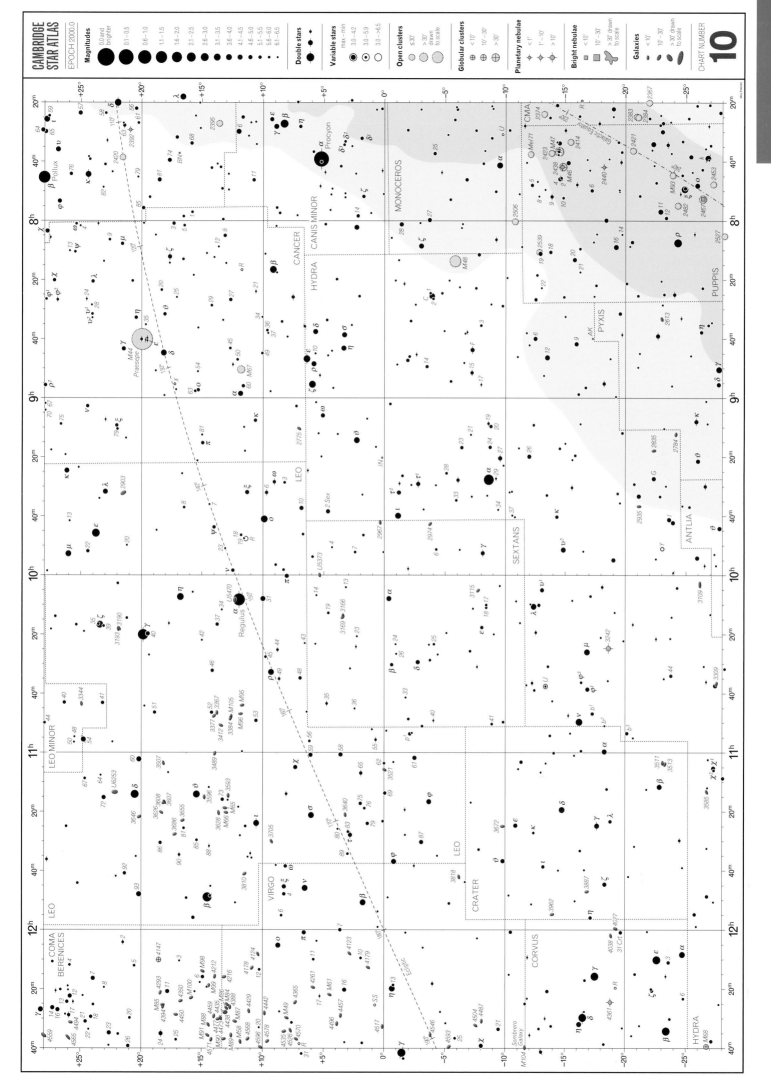

Chart 11 — RA 12ʰ to 16ʰ, declination +20° to −20°

Variable stars

Star	Con	RA h m	Dec ° '	Range	Type	Period (days)	Spectrum
R	Crv	12 19.6	−19 15	6.7–14.4	M	317.0	M
SS	Vir	12 25.3	+00 48	6.0–9.6	M	354.7	K
R	Vir	12 38.50	+06 59	6.0–12.1	M	145.6	M
S	Vir	13 33.0	−07 12	6.3–13.2	M	377.4	M
δ	Lib	15 01.0	−08 31	4.9–5.9	EA	2.33	B
R	Ser	15 50.7	+15 08	5.2–14.4	M	356.4	M

Double stars

Star	Con	RA h m	Dec ° '	PA °	Sep "	Magnitudes	
17	Vir	12 22.5	+05 18	337	20.0	6.6 + 9.4	
δ	Crv	12 29.9	−16 31	214	24.2	3.0 + 9.2	Algorab
24	Com	12 35.1	+18 23	271	20.3	5.2 + 6.7	
γ	Vir	12 41.7	−01 27	267	1.8	3.5 + 3.5	Porrima; binary, 171.4 years
ϑ	Vir	13 09.9	−05 32	343	7.1	4.4 + 9.4	
84	Vir	13 43.1	+03 32	229	2.9	5.5 + 7.9	
π	Boo	14 40.7	+16 25	108	5.6	4.9 + 5.8	
ζ	Boo	14 41.1	+13 44	300	0.8	4.5 + 4.6	Binary, 123.4 years
μ	Lib	14 49.3	−14 09	355	1.8	5.8 + 6.7	
α², α¹	Lib	14 50.9	−16 02	314	231.0	2.8 + 5.2	Zubenalgenubi
ι	Lib	15 12.2	−19 47	111	57.8	5.1 + 9.4	Binary, 22.4 years
δ	Ser	15 34.8	+10 32	176	4.4	4.2 + 5.2	Binary, 3168 years
47	Lib	15 55.0	−19 23	129	0.5	6.1 + 8.1	

Globular clusters

NGC/IC	Other	Con	RA h m	Dec ° '	Mag	Diam '
4147		Com	12 10.1	+18 33	10.3	4.0
5024	M53	Com	12 12.9	+18 10	7.7	12.6
5053		Com	13 16.4	+17 42	9.8	10.5
5634		Vir	14 29.6	−05 59	9.6	4.9
5904	M5	Ser	15 18.6	+02 05	5.8	17.4

Planetary nebula

NGC/IC	Other	Con	RA h m	Dec ° '	Mag p	Diam "	Mag*
4361		Crv	12 24.5	−18 48	10.3	45/110	13.2

Galaxies

NGC/IC	Other	Con	RA h m	Dec ° '	Mag	Size '	Type
4038		Vir	12 01.9	−18 52	10.7	2.6×1.8	Sc
4123		Vir	12 08.2	+02 53	11.2	4.50×3.5	SBb
4178		Vir	12 12.8	+10 52	11.4	5.0×2.0	SBc
4179		Vir	12 12.9	+01 18	10.9	4.2×1.2	S0
4192	M98	Com	12 13.8	+14 54	10.1	9.5×3.2	Sb
4212		Vir	12 15.7	+13 54	11.2	3.0×2.1	Sb
4216		Vir	12 15.9	+13 09	10.0	8.3×2.2	Sb
4254	M99	Com	12 18.8	+14 25	9.8	5.4×4.8	Sc
4261		Vir	12 19.4	+05 49	10.3	3.9×3.2	E2
4293		Com	12 21.2	+18 23	11.2	5.4×3.2	Sa
4303	M61	Vir	12 21.9	+04 28	9.7	6.0×5.5	Sc
4321	M100	Com	12 22.9	+15 49	9.4	6.9×6.2	Sc
4350		Vir	12 24.0	+16 42	11.1	3.2×1.1	E7
4365		Vir	12 24.5	+07 19	10.5	6.2×4.6	E7
4374	M84	Vir	12 25.1	+12 53	9.3	5.0×4.4	E1
4382	M85	Vir	12 25.4	+18 11	9.2	7.1×5.2	E
4388		Vir	12 25.8	+12 40	11.1	5.1×1.4	Sb
4394		Vir	12 25.9	+18 13	10.9	3.9×3.5	SBb
4406	M86	Vir	12 26.2	+12 57	9.2	7.4×5.5	E3
4429		Vir	12 27.4	+11 07	10.2	5.5×2.6	S0
4435		Vir	12 27.7	+13 05	10.9	3.0×1.0	E4
4438		Vir	12 27.8	+13 01	10.1	9.3×3.9	Sa
4442		Vir	12 28.1	+09 48	10.5	4.6×1.9	E5
4450		Com	12 28.5	+17 05	10.1	4.8×3.5	Sb
4457		Vir	12 29.0	+03 33	10.8	3.0×2.5	SBa
4459		Com	12 29.0	+13 59	10.4	3.8×2.8	E2
4472	M49	Vir	12 29.8	+08 00	8.4	8.9×7.4	E4
4473		Vir	12 29.8	+13 26	10.2	4.5×2.6	E4
4477		Com	12 30.0	+13 38	10.4	4.0×3.5	SBa
4486	M87	Vir	12 30.8	+12 24	8.6	7.2×6.8	E1
4487		Vir	12 31.1	−08 03	11.5	4.1×3.0	Sc
4496		Vir	12 31.7	+03 56	11.7	3.9×3.1	SBc
4501	M88	Com	12 32.0	+14 25	9.5	6.9×3.9	Sb
4504		Vir	12 32.3	−07 34	11.7	4.0×2.0	Sc
4517		Vir	12 32.8	+00 07	10.5	10.2×1.9	Sc
4526		Vir	12 34.0	+07 42	9.6	7.2×2.3	E7
4535		Vir	12 34.3	+08 12	9.8	6.8×5.0	SBc
4548	M91	Com	12 35.4	+14 30	10.2	5.4×4.4	SBb
4546		Vir	12 35.5	−03 48	10.3	3.5×1.7	E6
4552	M89	Vir	12 35.7	+12 33	9.8	4.2×4.2	E0
4568		Vir	12 36.6	+11 14	10.8	4.6×2.1	Sc
4569	M90	Vir	12 36.8	+13 10	9.5	9.5×4.7	Sb
4570		Vir	12 36.9	+07 15	10.9	4.1×1.3	S0
4571		Vir	12 36.9	+14 13	11.3	3.8×3.4	Sc
4578		Vir	12 37.5	+09 33	11.4	3.6×2.8	Sa
4579	M58	Vir	12 37.7	+11 49	9.8	5.4×4.4	Sb
4596		Vir	12 39.9	+10 11	10.5	3.9×2.8	SBa
4594	M104	Vir	12 40.0	−11 37	8.3	8.9×4.1	Sb
4621	M59	Vir	12 42.0	+11 39	9.8	5.1×3.4	E3
4636		Vir	12 42.8	+02 41	9.6	6.2×5.0	E1
4643		Vir	12 43.3	+01 59	10.6	3.4×2.7	SBa
4649	M60	Vir	12 43.7	+11 33	8.8	7.2×6.2	E1
4651		Com	12 44.0	+16 24	10.7	3.8×2.7	Sc
4654		Vir	12 44.0	+13 08	10.5	4.7×3.0	Sc
4665		Vir	12 45.1	+03 03	11.6	4.2×3.5	SB0
4666		Vir	12 45.1	−00 28	10.8	4.5×1.5	Sc
4689		Com	12 47.8	+13 46	10.9	4.0×3.5	Sc
4691		Vir	12 48.2	−03 20	11.2	3.2×2.7	SBb
4697		Vir	12 48.6	−05 48	9.3	6.0×3.8	E4
4698		Vir	12 48.4	+08 29	10.7	4.3×2.5	Sb
4699		Vir	12 49.0	−08 40	9.6	3.5×2.7	Sa
4710		Vir	12 49.6	+15 10	11.0	5.1×1.4	S0
4731		Com	12 51.0	−06 24	11.3	6.5×3.4	SBc
4754		Vir	12 52.3	+11 19	10.6	4.7×2.6	SB0
4753		Vir	12 52.4	−01 12	9.9	5.4×2.9	Pec
4762		Vir	12 52.9	+11 14	10.2	8.7×1.6	SB0
4781		Vir	12 54.4	−10 32	11.8	3.5×1.8	Sc
4818		Vir	12 56.8	−08 31	11.7	4.5×1.7	SB
4856		Vir	12 59.3	−15 02	10.4	4.6×1.6	SBa
4866		Vir	12 59.5	+14 10	11.0	6.5×1.5	Sb
4900		Vir	13 00.6	+02 30	11.5	2.3×2.2	Sc
4902		Vir	13 01.0	−14 31	11.2	3.0×2.8	Sb
4939		Vir	13 04.2	−10 20	11.4	5.8×3.2	Sb
4941		Vir	13 04.2	−05 33	11.1	3.7×2.1	Sb
4958		Vir	13 05.8	−08 01	10.5	4.1×1.4	E6
4984		Vir	13 09.0	−15 31	11.8	2.8×2.2	Sc
4995		Vir	13 09.7	−07 50	11.0	2.5×1.7	Sb
5018		Vir	13 13.0	−19 31	10.8	2.6×2.1	Sa
5044		Vir	13 15.4	−16 23	11.0	2.6×2.6	E0
5054		Vir	13 17.0	−16 38	11.3	5.0×3.1	Sb
5170		Vir	13 29.8	−17 58	11.8	8.1×1.3	E2
5247		Vir	13 38.1	−17 53	10.5	5.4×4.7	Sc
5248		Boo	13 37.5	+08 53	10.2	6.5×4.9	Sc
5363		Vir	13 56.1	+05 15	10.2	4.2×2.7	E
5364		Vir	13 56.2	+05 01	10.4	7.1×5.0	Sc
5427		Vir	14 03.4	−06 02	11.4	2.5×2.3	Sc
5566		Vir	14 20.3	+03 56	10.5	6.5×2.4	Sb
5576		Vir	14 21.1	+03 16	10.9	3.2×2.2	E2
5668		Vir	14 33.4	+04 27	11.5	3.3×3.1	Sc
5701		Vir	14 39.2	+05 22	11.8	4.7×4.5	SB
5713		Vir	14 40.2	−00 17	11.4	2.8×2.5	Sc
5746		Vir	14 44.9	+01 57	10.6	7.9×1.7	Sb
5813		Vir	15 01.2	+01 42	10.7	3.6×2.8	E1
5838		Vir	15 05.4	+02 06	10.8	4.2×1.6	Sa
5846		Vir	15 06.8	+01 36	10.2	3.4×3.2	E0
5885		Vir	15 15.1	−10 05	11.7	3.5×3.2	Sc

Notes: 4486 (M87) — Virgo A; 3C274. 4594 (M104) — Sombrero Galaxy.

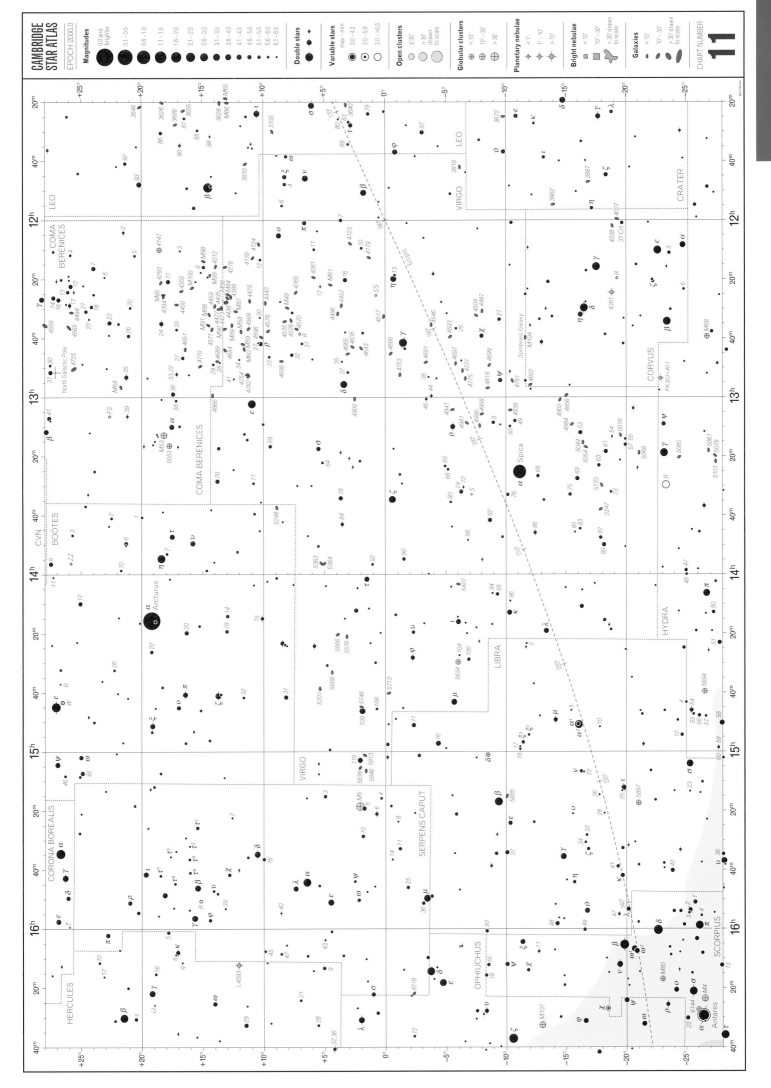

CAMBRIDGE
STAR ATLAS
EPOCH 2000.0

CHART NUMBER

11

Magnitudes

| 0.0 and brighter | 0.1–0.5 | 0.6–1.0 | 1.1–1.5 | 1.6–2.0 | 2.1–2.5 | 2.6–3.0 | 3.1–3.5 | 3.6–4.0 | 4.1–4.5 | 4.6–5.0 | 5.1–5.5 | 5.6–6.0 | 6.1–6.5 |

Double stars

Variable stars
max – min
3.0 – 4.2   3.0 – 5.9   3.0 – >6.5

Open clusters
≤30'   >30' drawn to scale

Globular clusters
<10'   10'–30'   >30'

Planetary nebulae
<1'   1'–10'   >10'

Bright nebulae
10'–30'   >30' drawn to scale

Galaxies
<10'   10'–30'   >30' drawn to scale

Chart 12　RA 16ʰ to 20ʰ, declination +20° to −20°

Variable stars

Star	Con	RA h m	Dec ° '	Range	Type	Period (days)	Spectrum
U	Her	16 25.8	+18 54	6.5–13.4	M	406.1	M
χ	Oph	16 27.0	−18 27	4.2–5.0	Irr	—	M
V1010	Oph	16 49.5	−15 40	6.1–7.0	EB	0.66	A
S	Her	16 51.9	+14 56	6.4–13.8	M	307.4	M
R	Oph	17 07.8	−16 06	7.0–13.8	M	302.6	M
α	Her	17 14.6	+14 23	2.7–4.0	SR	—	M
U	Oph	17 16.5	+01 13	5.9–6.6	EA	1.68	B
Y	Sgr	18 21.4	−18 52	5.4–6.1	Cep	5.77	F
59	Ser	18 27.2	+00 12	4.9–5.9	?	—	G + A
U	Sgr	18 31.9	−19 07	6.3–7.1	Cep	6.74	F – G
X	Oph	18 38.3	+08 50	5.9–9.2	M	334.4	M + K
V3879	Sgr	18 42.9	−19 17	6.1–6.6	SR	50:	M
R	Sct	18 47.5	−05 42	4.5–8.2	RV	140.1	G – K
FF	Aql	18 58.2	+17 22	5.2–5.7	Cep	4.47	F
V	Aql	19 04.4	−05 41	6.6–8.4	SR	353	K
R	Aql	19 06.4	+08 14	5.5–12.0	M	284.2	M
U	Sge	19 18.8	+19 37	6.6–9.2	EA	3.38	B + K
U	Aql	19 29.4	−07 03	6.1–6.9	Cep	7.02	F – G
η	Aql	19 52.5	+01 00	3.5–4.4	Cep	7.17	F – G
V505	Sgr	19 53.1	−14 36	6.5–7.5	EA	1.18	A + F
S	Sge	19 56.0	+16 38	5.3–6.0	Cep	8.38	F – G

Double stars

Star	Con	RA h m	Dec ° '	PA °	Sep "	Magnitudes	
β	Sco	16 05.4	−19 48	21	13.6	2.6 + 4.9	Graffias
11	Sco	16 07.6	−12 45	132	0.5	10.3	
κ	Her	16 08.1	+17 03	12	28.4	5.3 + 6.5	
ν	Sco	16 12.0	−19 28	337	41.1	4.3 + 6.8	
υ	Oph	16 27.8	−08 22	257	3.3	5.6 + 9.9	
λ	Oph	16 30.9	+01 59	30	1.5	4.2 + 5.2	Binary, 129.9 years
37, 36	Her	16 40.6	+04 13	230	69.8	5.8 + 7.0	
19	Oph	16 47.2	+02 04	89	23.4	6.1 + 9.4	
η	Oph	17 10.4	−15 43	237	0.6	3.0 + 3.5	Binary, 84.3 years
α	Her	17 14.6	+14 23	104	4.6	2.9 v + 5.4	Rasalgethi; var.; bin., 3,600 y.
41	Oph	17 16.6	−00 27	346	1.0	4.8 + 7.8	
ν	Ser	17 20.8	−12 51	28	46.3	4.3 + 8.3	
53	Oph	17 34.6	+09 35	191	41.2	5.8 + 8.5	
τ	Oph	18 03.1	−08 11	283	1.7	5.2 + 5.9	Binary, 280 years
70	Oph	18 05.5	+02 30	127	100.3	4.2 + 6.0	Binary, 88.1 years
59	Ser	18 27.2	+00 12	148	3.8	5.3 + 7.6	
θ	Ser	18 56.2	+04 12	318	22.3	4.5 + 5.4	
23	Aql	19 18.5	+01 05	5	3.1	5.3 + 9.3	
π	Aql	19 48.7	+11 49	110	1.4	6.1 + 6.9	
57	Aql	19 54.6	−08 14	170	35.7	5.8 + 6.5	

Open clusters

NGC/IC	Other	Con	RA h m	Dec ° '	Mag	Diam '	N*	
I.4665		Oph	17 46.3	+05 43	4.2	41	30	
6494	M23	Sgr	17 56.8	−19 01	5.5	27	150	
6595		Sgr	18 17.0	−19 53	7.0	11	30	
—	M24	Sgr	18 17:	−18 40:	—	120:	—	Star Cloud
6604		Ser	18 18.1	−12 14	6.5	2	30	
6613	M18	Sgr	18 19.9	−17 08	6.9	9	20	
6633		Oph	18 27.7	+06 34	4.6	27	30	
I.4725	M25	Sgr	18 31.6	−19 15	4.6	32	30	
6645		Sgr	18 32.6	−16 54	8.5	10	40	
I.4756		Ser	18 39.0	+05 27	5.4	52	80	
6694	M26	Sct	18 45.2	−09 24	8.0	15	30	
6705	M11	Sct	18 51.1	−06 16	5.8	14	500	Wild Duck Cluster
6709		Aql	18 51.5	+10 21	6.7	13	40	
6716		Sgr	18 54.6	−19 53	6.9	7	20	
6738		Aql	19 01.4	+11 36	8.3	15	100	
6755		Aql	19 07.8	+04 14	7.5	15	15	
—	H20	Sge	19 53.1	+18 20	7.7	7	—	

Globular clusters

NGC/IC	Other	Con	RA h m	Dec ° '	Mag	Diam '
6171	M107	Oph	16 32.5	−13 03	8.1	10.0
6218	M12	Oph	16 47.2	−01 57	6.6	14.5
6254	M10	Oph	16 57.1	−04 06	6.6	15.1
6333	M9	Oph	17 19.2	−18 31	7.9	9.3
6342		Oph	17 21.2	−19 35	9.9	3.0
6356		Oph	17 23.6	−17 49	8.4	7.2
6366		Oph	17 27.7	−05 05	10.0	8.3
6402	M14	Oph	17 37.6	−03 15	7.6	11.7
6517		Oph	18 01.8	−08 58	10.3	4.3
6535		Ser	18 03.8	−00 18	10.6	3.6
6539		Ser	18 04.8	−07 35	9.6	6.9
I.1276		Ser	18 10.7	−07 12	10.3	7.1
6712		Sct	18 53.1	−08 42	8.2	7.2
6760		Aql	19 11.2	+01 02	9.1	6.6
6838	M71	Sge	19 53.8	+18 47	8.3	7.2

Bright diffuse nebulae

NGC/IC	Other	Con	RA h m	Dec ° '	Type	Diam '	Mag*	
6611	M16	Ser	18 18.8	−13 47	E	35 × 28	—	Eagle Nebula, contains cluster
6618	M17	Sgr	18 20.8	−16 11	E	46 × 37	—	Omega Nebula, contains cluster

Planetary nebulae

NGC/IC	Other	Con	RA h m	Dec ° '	Mag p	Diam "	Mag*
I.4593		Her	16 12.2	+12 04	10.9	12/120	11.3
6309		Oph	17 14.1	−12 55	10.8	14/66	14.4
6537		Sgr	18 05.2	−19 51	12.5	9	—
6572		Oph	18 12.1	+06 51	9.0	8	13.6
6741		Aql	19 02.6	−00 27	10.8	6	14.7
6751		Aql	19 05.9	−06 00	12.5	20	13.9
6781		Aql	19 18.4	+06 33	11.8	109	15.0
6790		Aql	19 23.2	+01 31	10.2	7	13.5
6803		Aql	19 31.3	+10 03	11.3	6	15.2
—	PK52-2.2	Aql	19 39.2	+15 57	12.6	10	13.1
6818		Sgr	19 44.0	−14 09	9.9	17	

Galaxies

NGC/IC	Other	Con	RA h m	Dec ° '	Mag	Size	Type
6118		Ser	16 21.8	−02 17	12.3	4.7 × 2.3	Sb
6384		Oph	17 32.4	+07 04	10.6	6.0 × 4.3	Sb
6822		Sgr	19 44.9	−14 48	9.4	10.2 × 9.5	Irr

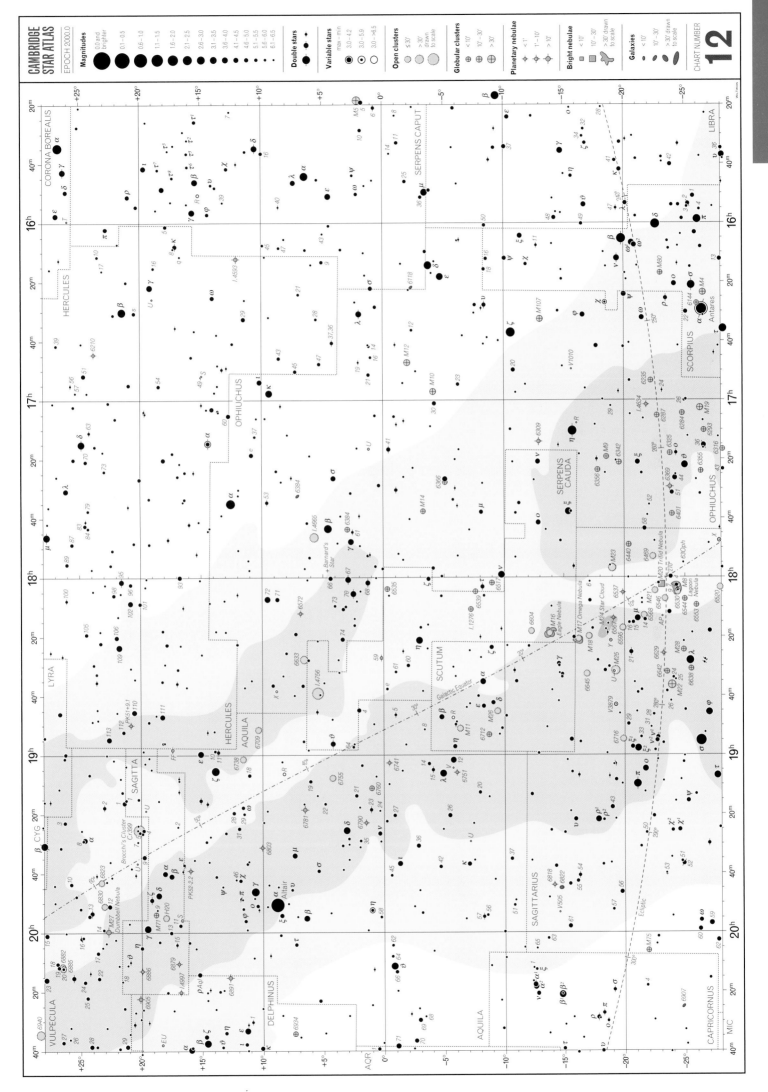

Chart 13 RA 20ʰ to 0ʰ, declination +20° to –20°

Variable stars

Star	Con	RA (h m)	Dec (° ')	Range	Type	Period (days)	Spectrum
EU	Del	20 37.9	+18 16	5.8–6.9	SR	59.5	M
U	Del	20 45.5	+15 05	5.6–7.5	SR	110:	M
EP	Aqr	21 46.5	–02 13	6.4–6.8	SR	55:	M
AG	Peg	21 51.0	+12 38	6.0–9.4	ZA	830.1	M
R	Aqr	23 43.8	–15 17	5.8–12.4	M	387.0	M
TX	Psc	23 46.4	+03 29	4.8–5.2	Irr	—	M

Double stars

Star	Con	RA (h m)	Dec (° ')	PA (°)	Sep (")	Magnitudes	
15	Sge	20 04.1	+17 04	276	190.7	5.9 + 9.1	
				320	203.7	6.8	
α¹	Cap	20 17.6	–12 30	221	45.4	4.2 + 9.2	
α²	Cap	20 18.1	–12 33	172	6.6	3.6 + 10.4	Algedi
α², α¹	Cap	20 18.1	–12 33	291	377.7	3.6 + 4.2	
σ	Cap	20 19.4	–19 07	179	55.9	5.5 + 9.0	
π	Cap	20 27.3	–18 13	148	3.2	5.3 + 8.9	
ρ	Cap	20 28.9	–17 49	158	0.5	5.0 + 10.0	
1	Del	20 30.3	+10 54	346	0.9	6.1 + 8.1	
γ	Del	20 46.7	+16 07	268	9.6	4.5 + 5.5	
13	Del	20 47.8	+06 00	194	1.6	5.6 + 9.2	
1	Eql	20 59.1	+04 18	284	0.8	6.0 + 6.3	= ε Eql; binary 101.4 years
γ	Eql	21 10.3	+10 08	268	1.9	4.7 + 11.5	
				5	47.7	12.5	
ζ	Aqr	22 28.8	–00 01	153	352.5	5.9	
				192	2.1	4.3 + 4.5	Binary, 856 years
37	Peg	22 30.0	+04 26	191	41.0	7.5	
				118	0.7	5.8 + 7.1	Binary, 140 years
107	Aqr	23 46.0	–18 41	136	6.6	5.7 + 6.7	

Open cluster

NGC/IC	Other	Con	RA (h m)	Dec (° ')	Mag	Diam (')	N*	
6994	M73	Cep	20 59.0	–12 38	8.9	2.8	4	Not a real cluster

Globular clusters

NGC/IC	Other	Con	RA (h m)	Dec (° ')	Mag	Diam (')
6934		Del	20 34.2	+07 24	8.9	5.9
6981	M72	Aqr	20 53.5	–12 32	9.4	5.9
7006		Del	21 01.5	+16 11	10.6	2.8
7078	M15	Peg	21 30.0	+12 10	6.4	12.3
7089	M2	Aqr	21 33.5	–00 49	6.5	12.9
7492		Aqr	23 08.4	–15 37	11.5	6.2

Planetary nebulae

NGC/IC	Other	Con	RA (h m)	Dec (° ')	Mag (p)	Diam (")	Mag*	
6879		Sge	20 10.5	+16 55	13.0	5	15:	
6886		Sge	20 12.7	+19 59	12.2	4	15.7	
6891		Del	20 15.2	+12 42	11.7	12/74	12.4	
I.4997		Sge	20 20.2	+16 45	11.6	2	13:	
7009		Aqr	21 04.2	–11 22	8.3	25/100	11.5	Saturn Nebula

Galaxies

NGC/IC	Other	Con	RA (h m)	Dec (° ')	Mag	Size	Type
7448		Peg	23 00.1	+15 59	11.7	2.7 × 1.3	Sc
7479		Peg	23 04.9	+12 19	11.0	4.1 × 3.2	SBb
7606		Aqr	23 19.1	–08 29	10.8	5.8 × 2.6	Sb
7723		Aqr	23 38.9	–12 58	11.1	3.6 × 2.6	Sb
7727		Aqr	23 39.9	–12 18	10.7	4.2 × 3.4	SBa

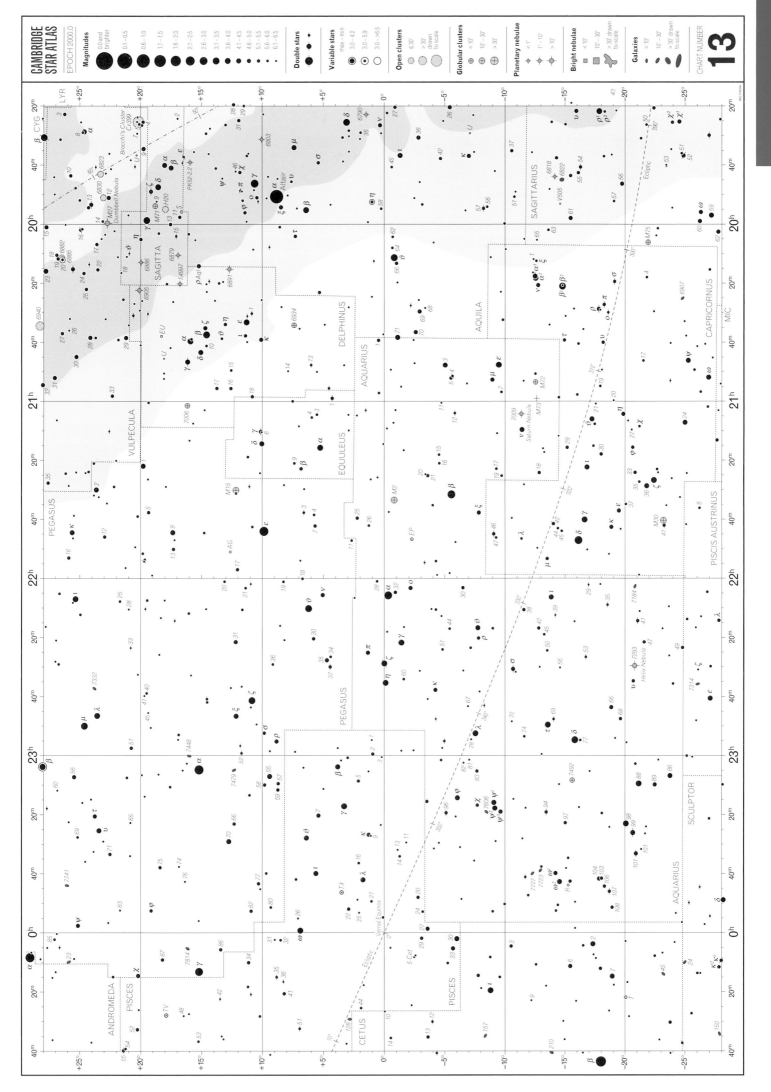

CAMBRIDGE STAR ATLAS

EPOCH 2000.0

CHART NUMBER

13

Chart 14　RA 0ʰ to 4ʰ, declination –20° to –65°

Variable stars

Star	Con	RA h m	Dec ° '	Range	Type	Period (days)	Spectrum
S	Scl	00 15.4	–32 03	5.5–13.6	M	365.3	M
T	Cet	00 21.8	–20 03	5.0–6.9	SR	158.9	M
ζ	Phe	01 08.4	–55 15	3.9–4.4	EA	1.67	B
R	Scl	01 27.0	–32 33	9.1–12.8 p	SR	370	M
AA	Cet	01 59.0	–22 55	6.0–6.5	EW	0.54	F
R	Hor	02 53.9	–49 53	4.7–14.3	M	404.0	M
V	Hor	03 03.5	–58 56	8.7–9.8 p	SR	—	M
TW	Hor	03 12.6	–57 19	5.3–6.0	SR	158:	K

Double stars

Star	Con	RA h m	Dec ° '	PA °	Sep "	Magnitudes	Spectrum
κ¹	Scl	00 09.3	–27 59	265	1.4	6.1 + 6.2	
β	Tuc	00 31.5	–62 58	169	27.1	4.4 + 4.8	
ξ	Phe	00 41.8	–56 30	253	13.2	5.8 + 10.2	
λ²	Scl	00 42.7	–38 28	3	0.5	6.7 + 7.0	
β	Phe	01 06.1	–46 43	346	1.4	4.0 + 4.2	
p	Eri	01 39.8	–56 12	191	11.5	5.8 + 5.8	Binary, 483.7 years
ε	Scl	01 45.6	–25 03	23	4.7	5.4 + 8.6	Binary, 1192 years
ω	For	02 33.8	–28 14	244	10.8	5.0 + 7.7	
η²	For	02 50.2	–35 51	14	5.0	5.9 + 10.1	
ϑ	Eri	02 58.3	–40 18	88	8.2	3.4 + 4.5	
α	For	03 12.1	–28 59	299	5.1	4.0 + 7.0 v	Binary, 314 years
τ⁴	Eri	03 19.5	–21 45	288	5.7	3.7 + 9.2	
				236	160.2	9.7	
χ³	For	03 28.2	–35 51	248	6.3	6.5 + 10.5	

Globular clusters

NGC/IC	Other	Con	RA h m	Dec ° '	Mag	Diam '
288		Scl	00 52.8	–26 35	8.1	13.8
1261		Hor	03 12.3	–55 13	8.4	6.9

Planetary nebula

NGC/IC	Other	Con	RA h m	Dec ° '	Mag p	Diam "	Mag*
1360		For	03 33.3	–25 51	9.4	390	11.4

Galaxies

NGC/IC	Other	Con	RA h m	Dec ° '	Mag	Size '	Type	
24		Scl	00 09.9	–24 58	11.5	5.5 × 1.6	Sb	
45		Cet	00 14.1	–23 11	10.4	8.1 × 5.8	S	
55		Scl	00 14.9	–39 11	8.2	32.4 × 6.5	SB	
134		Scl	00 30.4	–33 15	10.1	8.1 × 2.6	SBb	
150		Scl	00 34.3	–27 48	11.1	4.2 × 2.3	S	
247		Scl	00 47.1	–20 46	8.9	20.0 × 7.4	Sc	
253		Scl	00 47.6	–25 17	7.1	25.1 × 7.4	Sc	
289		Scl	00 52.7	–31 12	11.6	3.7 × 2.7	Sb	
300		Scl	00 54.9	–37 41	8.7	20.0 × 14.8	Sd	
—		Scl	00 59.9	–33 42	10.5	35: × 29:	dE3	Sculptor Dwarf
578	E 351–30	Cet	01 30.5	–22 40	10.9	4.8 × 3.2	Sc	
613		Scl	01 34.3	–29 25	10.0	5.8 × 4.6	SBc	
685		Eri	01 47.8	–52 47	11.8	4.1 × 4.0	SBb	
908		Cet	02 23.1	–21 14	10.2	5.5 × 2.8	SBb	
986		For	02 33.6	–39 02	11.0	3.7 × 2.8	Sc	
—	E 356–4	For	02 39.9	–34 32	9.0	63: × 48:	dE3	Fornax Dwarf
1097		Eri	02 46.3	–30 17	9.3	9.3 × 6.6	SBb	
1187		Eri	03 02.6	–22 52	10.9	5.0 × 4.0	SBc	
1201		For	03 04.1	–26 04	10.6	4.4 × 2.8	Sa	
1232		Eri	03 09.8	–20 35	9.9	7.8 × 6.9	Sc	
1249		Hor	03 10.1	–53 21	11.7	5.2 × 2.7	SBc	
1255		For	03 13.5	–25 44	11.1	4.1 × 2.8	Sc	
1291		Eri	03 17.3	–41 08	8.5	10.5 × 9.1	SBa	
1302		For	03 19.9	–26 04	11.5	4.4 × 4.2	SBa	
1316		For	03 22.7	–37 12	8.9	7.1 × 5.5	SBa	
1326		For	03 23.9	–36 28	10.5	4.0 × 3.0	SB0	
1332		Eri	03 26.3	–21 20	10.3	4.6 × 1.7	E7	
1344		For	03 28.3	–31 04	10.3	3.9 × 2.3	E3	
1350		For	03 31.1	–33 38	10.5	4.3 × 2.4	SBb	
1365		For	03 33.6	–36 08	9.5	9.8 × 5.5	SBb	
1371		For	03 35.0	–24 56	11.5	5.4 × 4.0	SBa	
1380		For	03 36.5	–34 59	11.1	4.9 × 1.9	S0	
1385		For	03 37.5	–24 30	11.2	3.0 × 2.0	Sc	
1395		Eri	03 38.5	–23 02	11.3	3.2 × 2.5	E3	
1398		For	03 38.9	–26 20	9.7	6.6 × 5.2	SBb	
1399		For	03 38.5	–35 27	9.9	3.2 × 3.1	E1p	
1404		For	03 38.8	–44 05	11.9	2.8 × 2.3	S0	
1411		Hor	03 38.8	–35 35	10.3	2.5 × 2.3	E1	
1433		Hor	03 42.0	–47 13	10.0	6.8 × 6.0	SBa	
1425		For	03 42.2	–29 54	11.7	5.4 × 2.7	Sb	
1448		Hor	03 44.5	–44 39	11.3	8.1 × 1.8	Sc	
1493		Hor	03 57.5	–46 12	11.8	2.6 × 2.3	SBc	

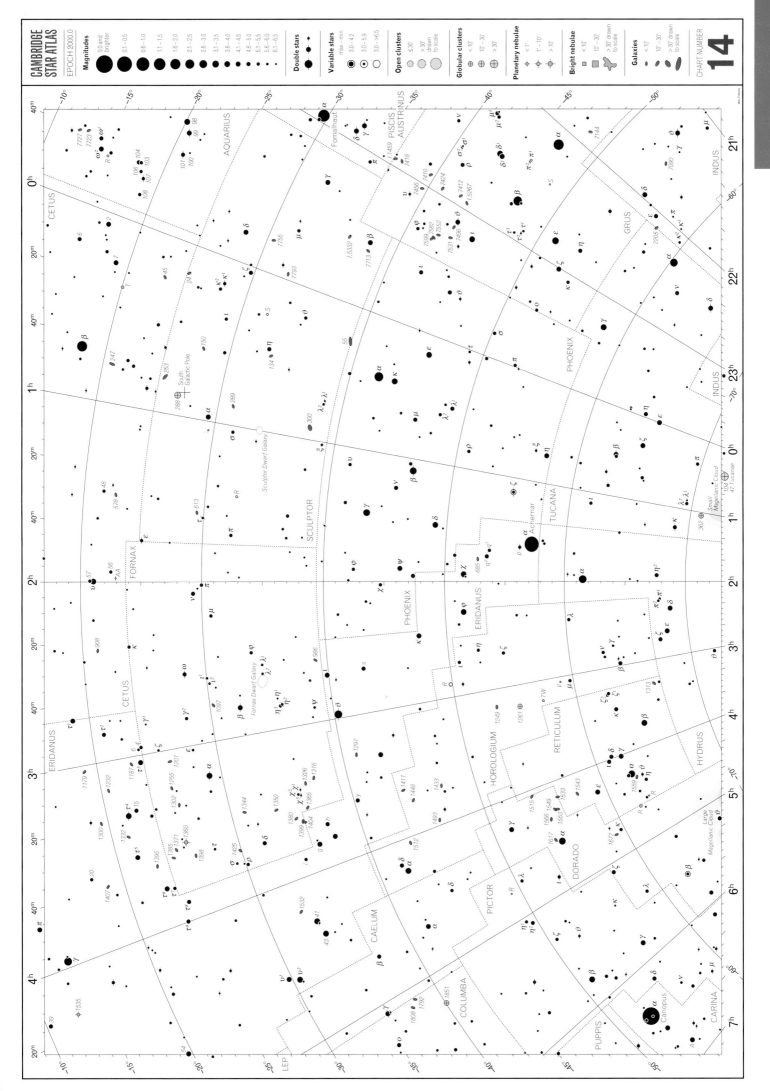

CAMBRIDGE STAR ATLAS

EPOCH 2000.0

Magnitudes

0.0 and brighter
0.1–0.5
0.6–1.0
1.1–1.5
1.6–2.0
2.1–2.5
2.6–3.0
3.1–3.5
3.6–4.0
4.1–4.5
4.6–5.0
5.1–5.5
5.6–6.0
6.1–6.5

Double stars

Variable stars

max – min
3.0 – 4.2
3.0 – 5.9
3.0 – >6.5

Open clusters

≤30'
>30' drawn to scale

Globular clusters

<10'
10'–30'
>30'

Planetary nebulae

<1'
1'–10'
>10'

Bright nebulae

<10'
10'–30'
>30' drawn to scale

Galaxies

<10'
10'–30'
>30' drawn to scale

CHART NUMBER

14

Chart 15 RA 4ʰ to 8ʰ, declination −20° to −65°

Variable stars

Star	Con	RA h m	Dec ° '	Range	Type	Period (days)	Spectrum
R	Ret	04 33.5	−63 02	6.5–14.0	M	278.3	M
R	Dor	04 36.8	−62 05	4.8–6.6	SR	338:	M
R	Pic	04 46.2	−49 15	6.7–10.0	SR	164.2	M
β	Dor	05 33.6	−62 29	3.5–4.1	Cep	9.84	F–G
S	Lep	06 05.8	−24 12	6.0–7.6	SR	90	B
L²	Pup	07 13.5	−44 39	2.6–6.2	SR	140.4	M
ω	CMa	07 14.8	−26 46	3.6–4.2	Irr	—	B
V	Pup	07 58.2	−49 15	4.4–4.9	EB	1.45	B+B

Double stars

Star	Con	RA h m	Dec ° '	PA °	Sep "	Magnitudes	
γ	Cae	05 40.4	−25 29	308	2.9	4.6 + 8.1	
θ	Pic	05 24.8	−52 19	287	38.2	6.3 + 6.8	
β	Lep	05 28.2	−20 46	330	2.5	2.8 + 7.3	Nihal
γ	Lep	05 44.5	−22 27	350	96.3	3.7 + 6.3	
μ	Pic	06 32.0	−58 45	231	2.4	5.8 + 9.0	
17	CMa	06 55.0	−20 24	147	44.4	5.8 + 9.3	
π	CMa	06 55.6	−20 08	18	11.6	4.7 + 9.7	
ε	CMa	06 58.6	−28 58	161	7.5	1.5 + 7.4	Adhara
σ	Pup	07 29.2	−43 18	74	22.3	3.3 + 9.4	

Open clusters

NGC/IC	Other	Con	RA h m	Dec ° '	Mag	Diam '	N*	
2287	M41	CMa	06 47.0	−20 44	4.5	38	80	
2354		CMa	07 14.3	−25 44	6.5	20	100	
2362		CMa	07 18.8	−24 57	4.1	8	60	
—	Cr 135	Pup	07 17.0	−36 50	2.1	50	—	Contains π Pup
2367		CMa	07 20.1	−21 56	7.9	3.5	30	
2383		CMa	07 24.8	−20 56	8.4	6	40	
2384		CMa	07 25.1	−21 02	7.4	2.5	15	
2421		Pup	07 36.3	−20 37	8.3	10	70	
2439		Pup	07 40.8	−31 39	6.9	10	80	
2447	M93	Pup	07 44.6	−23 52	6.2	22	80	
2451		Pup	07 45.4	−37 58	2.8	45	40	Contains c Pup
2453		Pup	07 47.8	−27 14	8.3	5	30	
2477		Pup	07 52.3	−38 33	5.8	27	160	
2467		Pup	07 52.6	−26 23	7.1	16	50	Contains nebula
2482		Pup	07 54.9	−24 18	7.3	12	40	
2489		Pup	07 56.2	−30 04	7.9	8	45	
2516		Car	07 58.3	−60 53	3.8	30	80	

Globular clusters

NGC/IC	Other	Con	RA h m	Dec ° '	Mag	Diam '
1851		Col	05 14.1	−40 03	7.3	11.0
1904	M79	Lep	05 24.5	−24 33	8.0	8.7
2298		Pup	06 49.0	−36 00	9.4	6.8

Bright diffuse nebula

NGC/IC	Other	Con	RA h m	Dec ° '	Type	Diam '	Mag*	
2467		Pup	07 52.6	−26 24	E	8×7	9.2	In cluster

Galaxies

NGC/IC	Other	Con	RA h m	Dec ° '	Mag	Size '	Type
1512		Hor	04 03.9	−43 21	10.6	4.0×3.2	SBa
1515		Dor	04 04.1	−54 06	11.0	5.4×1.3	SBb
1532		Eri	04 12.1	−32 52	11.1	5.6×1.8	Sb
1543		Ret	04 12.8	−57 44	10.6	3.9×2.1	SB0
1533		Dor	04 16.2	−55 47	9.5	4.1×2.8	S0
1549		Dor	04 15.7	−55 36	9.9	3.7×3.2	E0
1553		Dor	04 16.2	−55 47	9.5	4.1×2.3	S0
1559		Ret	04 17.6	−62 47	10.5	3.3×2.1	SBc
1566		Dor	04 20.0	−54 56	9.4	7.6×6.2	SBb
1617		Dor	04 31.7	−54 36	10.4	4.7×2.4	SBa
1672		Dor	04 45.7	−59 15	11.0	4.8×3.9	SBb
1744		Lep	05 00.0	−26 01	11.2	6.8×4.1	SBc
1792		Col	05 05.2	−37 59	10.2	4.0×2.1	Sb
1808		Col	05 07.7	−37 31	9.9	7.2×4.1	SBa
1964		Lep	05 33.4	−21 57	10.8	6.2×2.5	Sb
2090		Col	05 47.0	−34 14	11.8	4.5×2.3	Sc
2207		CMa	06 16.4	−21 22	10.7	4.3×2.9	Sc
2217		CMa	06 21.7	−27 14	10.4	4.8×4.4	SBa
2223		CMa	06 24.6	−22 50	11.4	3.3×3.0	SBb
2280		CMa	06 44.8	−27 38	11.8	5.6×3.2	Sb

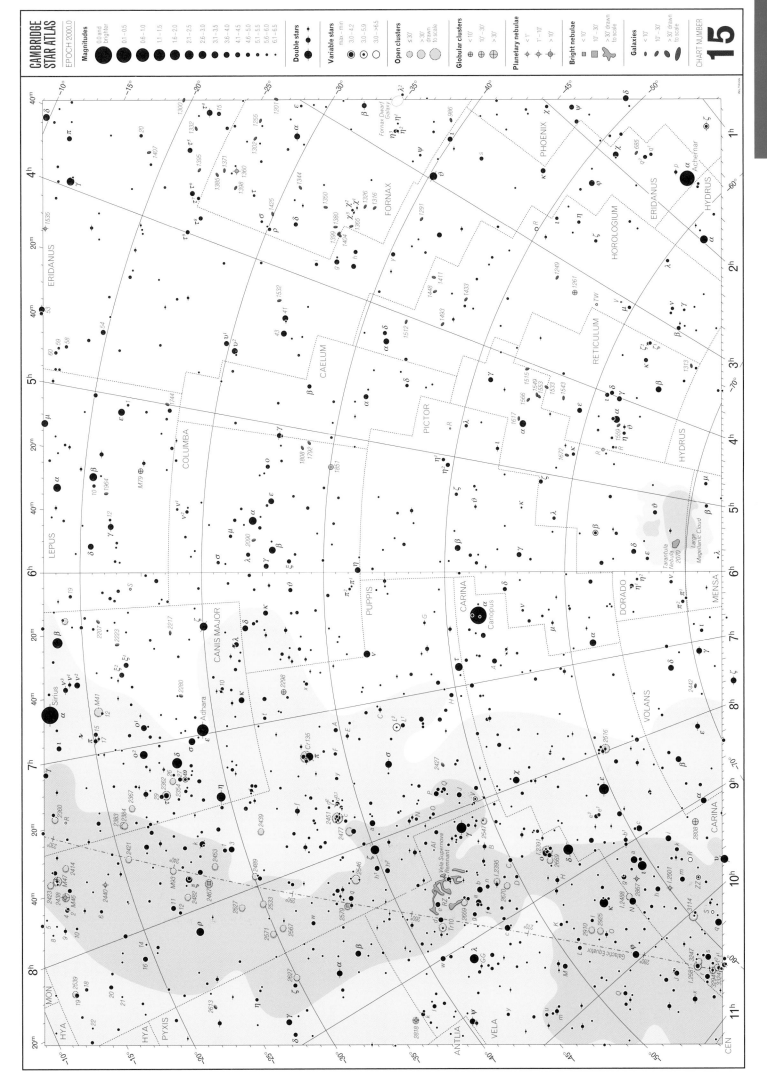

CAMBRIDGE STAR ATLAS

EPOCH 2000.0

Magnitudes

0.0 and brighter
0.1 – 0.5
0.6 – 1.0
1.1 – 1.5
1.6 – 2.0
2.1 – 2.5
2.6 – 3.0
3.1 – 3.5
3.6 – 4.0
4.1 – 4.5
4.6 – 5.0
5.1 – 5.5
5.6 – 6.0
6.1 – 6.5

Double stars

Variable stars
max – min
3.0 – 4.2
3.0 – 5.9
3.0 – >6.5

Open clusters
<30'
>30' drawn to scale

Globular clusters
<10'
10' – 30'
>30'

Planetary nebulae
<1'
1' – 10'
>10'

Bright nebulae
<10'
10' – 30'
>30' drawn to scale

Galaxies
<10'
10' – 30'
>30' drawn to scale

CHART NUMBER

15

Chart 16 RA 8ʰ to 12ʰ, declination −20° to −65°

Variable stars

Star	Con	RA (h m)	Dec (° ')	Range	Type	Period (days)	Spectrum
Al	Vel	08 14.1	−44 34	6.4–7.1	δSct	0.11	A–F
RZ	Vel	08 37.0	−44 07	6.4–7.6	Cep	20.39	G
R	Car	09 32.2	−62 47	3.9–10.5	M	308.7	M
ZZ	Car	09 45.2	−62 30	3.3–4.2	Cep	35.53	F–K
Y	Hya	09 51.1	−23 01	8.3–12.0 p	SR	302.8	K
S	Car	10 09.4	−61 33	4.5–9.9	M	149.5	K–M
U	Ant	10 35.2	−39 34	8.1–9.7 p	Irr	—	K
η	Car	10 45.1	−59 41	−0.8–7.9	SD	—	Pec
U	Car	10 57.8	−59 44	5.7–7.0	Cep	38.76	F–G
ER	Car	11 09.7	−58 50	6.6–7.1	Cep	7.72	F–G
o'	Cen	11 31.8	−59 27	4.7–5.5	SR	200:	G

Double stars

Star	Con	RA (h m)	Dec (° ')	PA (°)	Sep (")	Magnitudes
γ	Vel	08 09.5	−47 20	220	41.2	1.9 + 4.2
				151	62.3	8.3
				141	93.5	9.1
δ	Vel	08 44.7	−54 43	146	1.8	2.1 + 5.1
	Vel	08 56.3	−52 43	153	2.6	4.8 + 7.4
κ	Pyx	09 08.0	−25 52	339	2.7	4.6 + 9.8
ζ	Ant	09 30.8	−31 53	263	2.1	6.2 + 7.1
δ	Ant	09 29.6	−30 36	212	8.0	5.6 + 9.6
μ	Vel	10 46.8	−49 25	226	11.0	2.7 + 6.4
				58	2.5	
β	Hya	11 52.9	−33 54	8	0.9	4.7 + 5.5

Note: Binary, 116.2 years

Open clusters

NGC/IC	Other	Con	RA (h m)	Dec (° ')	Diam (')	Mag	N*	Notes
2527		Pup	08 05.3	−28 10	22	6.5	40	
2533		Pup	08 07.0	−29 54	3.5	7.6	60	
2547		Vel	08 10.7	−49 16	20	4.7	80	
2546		Pup	08 12.4	−37 38	41	6.3	40	
2567		Pup	08 18.6	−30 38	10	7.4	40	
2571		Pup	08 18.9	−29 44	13	7.0	30	
2579		Pup	08 21.1	−36 11	10	7.5	20	
2627		Pyx	08 37.3	−29 57	11	8.4	60	
I.2391		Vel	08 40.2	−53 04	50	2.5	30	o Velorum cluster
I.2395		Vel	08 41.1	−48 12	8	4.6	40	
2669		Vel	08 44.9	−52 58	12	6.1	40	
2670	Tr 10	Vel	08 45.5	−48 47	9	7.8	30	
—		Vel	08 47.8	−42 29	15	4.6	40	
2818		Pyx	09 16.0	−36 37	9	8.2	40	Contains planetary
I.2488		Vel	09 27.6	−56 59	15	7.4	70	
2910		Vel	09 30.4	−52 54	5	7.2	30	
2925		Vel	09 33.7	−53 26	12	8.3	40	
3114		Car	10 02.7	−60 07	35	4.2	100	
3247		Car	10 25.9	−57 56	7	7.6	20	
I.2581		Car	10 27.4	−57 38	8	4.3	25	
3293		Car	10 35.8	−58 14	6	4.7	30	
3324		Car	10 37.3	−58 38	6	6.7	10	
I.2602		Cep	10 43.2	−64 24	50	1.9	60	Southern Pleiades
3532		Car	11 06.4	−58 40	55	3.0	150	
3572		Car	11 10.4	−60 14	7	6.6	35	
3590		Car	11 12.9	−60 47	4	8.2	25	
I.2714		Car	11 17.9	−62 42	12	8.2	100	
—	Mel 105	Car	11 19.5	−63 30	4		70	
3680		Cen	11 25.7	−43 15	12	7.6	30	
3766		Cen	11 36.1	−61 37	12	5.3	100	
I.2944		Cen	11 36.6	−63 02	15	4.5	30	λ Centauri cluster
3960		Cen	11 50.9	−55 42	7	8.3	45	

Globular clusters

NGC/IC	Other	Con	RA (h m)	Dec (° ')	Mag	Diam (')
2808		Car	09 12.0	−64 52	6.3	13.8
3201		Vel	10 17.6	−46 25	6.8	18.2

Bright diffuse nebulae

NGC/IC	Other	Con	RA (h m)	Dec (° ')	Type	Diam (')	Mag*	Notes
2579		Pup	08 20.9	−36 13	E	2	1.2	
—	Gum 12	Vel	08 30:	−45:	E	1200×720		Vela SNR
3372		Car	10 43.8	−59 52	E	120×120	6.2	Eta Carinae Nebula

Planetary nebulae

NGC/IC	Other	Con	RA (h m)	Dec (° ')	Mag p	Diam (")	Mag*	Notes
2818		Pyx	09 16.0	−36 38	13.0	38		In open cluster
2867		Car	09 21.4	−58 19	9.7	11	13.6	
I.2501		Car	09 38.8	−60 05	11.3	25		
3132		Vel	10 07.7	−40 26	8.2	47	10.1	
3211		Car	10 17.8	−62 40	11.8	12		
3918		Cen	11 50.3	−57 11	8.4	12	10.8	

Galaxies

NGC/IC	Other	Con	RA (h m)	Dec (° ')	Mag	Size (')	Type
2613		Pyx	08 33.4	−22 58	10.4	7.2 × 2.1	Sb
2784		Hya	09 12.3	−24 10	10.1	5.1 × 2.3	S0
2835		Hya	09 17.9	−22 21	11.1	6.3 × 4.4	S
2935		Hya	09 36.7	−21 08	11.9	3.5 × 3.0	SBb
2997		Ant	09 45.6	−31 11	10.6	8.1 × 6.5	Sc
3109		Hya	10 03.1	−26 09	10.4	14.5 × 3.5	Irr
3223		Ant	10 21.6	−34 16	11.8	4.1 × 2.6	Sb
3256		Vel	10 27.8	−43 54	11.3	3.5 × 2.0	Pec
3309		Hya	10 36.6	−27 31	11.9	1.9 × 1.7	E0
3511		Crt	11 03.4	−23 05	11.6	5.4 × 2.2	Sc
3513		Crt	11 03.8	−23 15	12.0	2.8 × 2.3	SBc
3557		Cen	11 10.0	−37 32	10.4	4.0 × 2.7	E3
3585		Hya	11 13.3	−26 45	10.0	2.9 × 1.6	E5
3621		Hya	11 18.3	−32 49	9.9	10.0 × 6.5	Sc
3904		Hya	11 49.2	−29 17	11.0	2.2 × 1.7	E2
3923		Hya	11 51.0	−28 48	10.1	2.9 × 1.9	E3

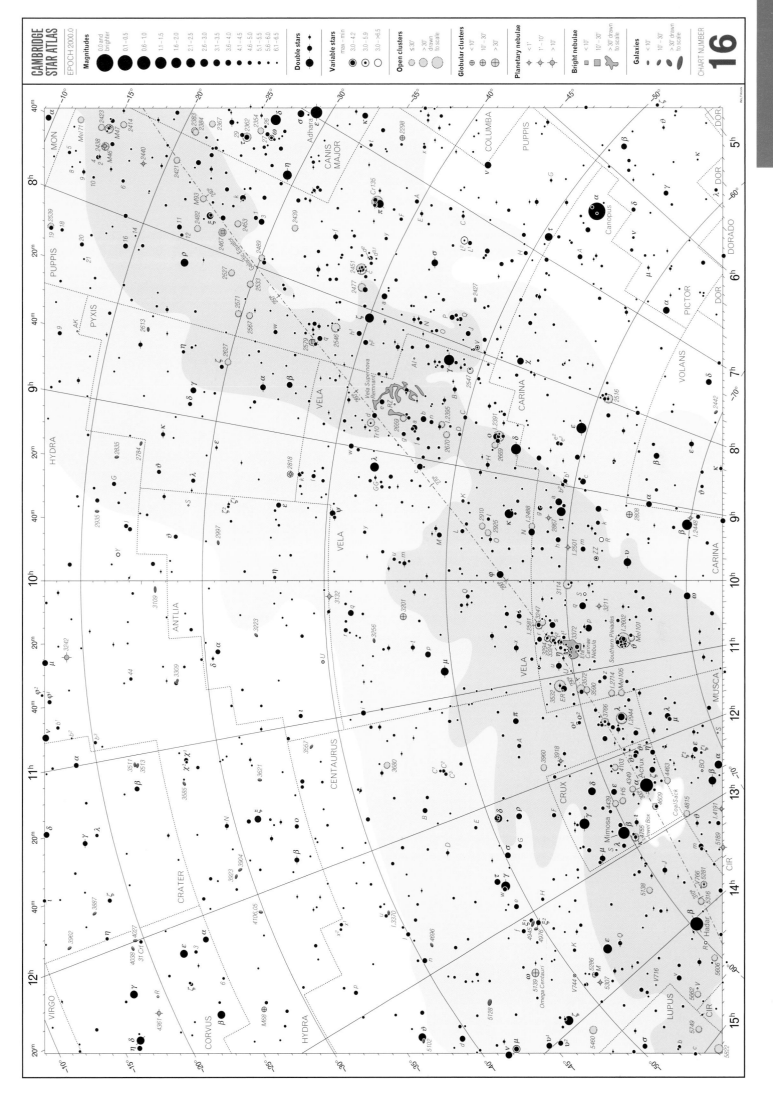

Chart 17 RA 12ʰ to 16ʰ, declination −20° to −65°

Variable stars

Star	Con	RA h m	Dec ° '	Range	Type	Period (days)	Spectrum
S	Cru	12 54.4	−58 26	6.2–6.9	Cep	4.69	F–G
R	Hya	13 29.7	−23 17	3.5–10.9	M	389.6	M
V744	Cen	13 40.0	−49 57	5.1–6.6	SR	90:	M
T	Cen	13 41.8	−33 36	5.5–9.0	SR	90.44	K–M
V766	Cir	13 47.2	−62 35	6.2–7.5	SD?	—	G
μ	Cen	13 49.6	−42 28	2.9–3.5	Irr	—	B
V716	Cen	14 13.7	−54 38	6.0–6.5	EB	1.49	B
R	Cen	14 16.6	−59 55	5.3–11.8	M	546.2	M
GG	Lup	15 32.2	−23 53	5.5–6.0	SR	—	B
R	Nor	15 36.0	−49 30	6.5–13.9	SR	492.7	M
T	Nor	15 44.1	−54 59	6.2–13.6	M	242.6	M

Double stars

Star	Con	RA h m	Dec ° '	PA °	Sep "	Magnitudes	
α	Cru	12 26.6	−63 06	115	4.4	1.4+1.9	Acrux
				202	90.1	4.9	
γ	Cru	12 31.2	−57 07	31	110.6	1.6+6.7	Gacrux
				82	155.2	9.5	
γ	Cen	12 41.5	−48 58	347	1.0	2.9+2.9	Binary, 84.5 years
ι	Cru	12 45.6	−60 59	22	26.9	4.7+9.5	
μ	Cru	12 54.6	−57 11	17	34.9	4.3+5.3	
3	Cen	13 51.8	−33 00	108	7.9	4.5+6.0	
4	Cen	13 53.2	−31 56	185	14.9	4.7+8.4	
β	Cen	14 03.8	−60 22	251	1.3	0.7+3.9	Hadar
α	Cen	14 39.6	−60 50	222	14.1	0.0+1.2	Rigil Kent; binary, 79.9 years
				211:	7860(131')	11.0	Proxima Centauri
α	Cir	14 42.5	−64 59	232	15.7	3.2+8.6	
54	Hya	14 46.0	−25 27	126	8.6	5.1+7.1	
59	Hya	14 58.7	−27 39	80	0.3	6.3+6.6	Binary, 339.3 years
π	Lup	15 05.1	−47 03	73	1.4	4.6+4.7	
κ	Lup	15 11.9	−48 44	144	26.8	3.9+5.8	
μ	Lup	15 18.5	−47 53	142	1.2	5.1+5.2	
γ	Cir	15 23.4	−59 19	20	0.7	5.1+5.5	Binary, 180 years
2	Sco	15 53.6	−25 20	274	2.5	4.7+7.4	
ξ	Lup	15 56.9	−33 58	49	10.4	5.3+5.8	

Globular clusters

NGC/IC	Other	Con	RA h m	Dec ° '	Mag	Diam '	
4590	M68	Hya	12 39.5	−26 45	8.2	12.0	
5139	ω Cen	Cen	13 26.8	−47 29	3.7	36.3	Omega Centauri
5286		Cen	13 46.4	−51 22	7.6	9.1	
5824		Lup	15 04.0	−33 04	9.0	6.2	
5897		Lib	15 17.4	−21 01	8.6	12.6	
5927		Lup	15 28.0	−50 40	8.3	12.0	
5946		Nor	15 35.5	−50 40	9.6	7.1	
5986		Lup	15 46.1	−37 47	7.1	9.8	

Nebulae

NGC/IC	Other	Con	RA h m	Dec ° '	Type	Diam '	
	Coal Sack	Cru	12 53:	−63:	—	400×300	Dark nebula
5367		Cen	13 57.7	−39 59	R	4×3	

Planetary nebulae

NGC/IC	Other	Con	RA h m	Dec ° '	Mag p	Diam "	Mag*
—	PK303+40.1	Hya	12 53.6	−22 52	12.0	709	9.6
5307		Cen	13 51.1	−51 12	12.1	13	
I.4406		Lup	14 22.4	−44 09	10.6	28	14.7
5873		Lup	15 12.8	−38 08	13.3	3	13.6
5882		Lup	15 16.8	−45 39	10.5	7	12.0

Galaxies

NGC/IC	Other	Con	RA h m	Dec ° '	Mag	Size	Type
4105		Hya	12 06.7	−29 46	12.0	2.4×1.9	E2
4106		Hya	12 06.8	−29 46	11.4	1.9×1.5	E0
I.3370		Cen	12 27.6	−39 20	11.1	2.8×2.4	E2
4696		Cen	12 48.8	−41 19	10.7	3.5×3.2	E1
4945		Cen	13 05.4	−49 28	9.5	20.0×4.4	SBc
4976		Cen	13 08.6	−49 30	10.2	4.3×2.6	E4
5061		Cen	13 18.1	−26 50	11.7	2.6×2.3	E2
5068		Vir	13 18.9	−21 02	10.8	6.9×6.3	SBc
5078		Hya	13 19.8	−27 24	12:	3.2×1.7	Sa
5085		Hya	13 20.3	−24 26	11.9	3.4×3.0	Sb
5101		Hya	13 21.8	−27 26	11.7	5.5×4.9	SBa
5102		Cen	13 22.0	−36 38	9.7	9.3×3.5	S0
5128		Cen	13 25.5	−43 01	7.0	18.2×14.5	S0
I.4296		Cen	13 36.6	−33 58	11.6	2.7×2.7	E0
5236	M83	Hya	13 37.0	−29 52	8.2	11.2×10.2	Sc
5253		Cen	13 39.9	−31 39	10.6	4.0×1.7	E5
5483		Cen	14 10.4	−43 19	12.0	3.1×2.8	Sc
5530		Lup	14 18.5	−43 24	11.9	4.1×2.2	Sb
5643		Lup	14 32.7	−44 10	10.7	4.6×4.1	SBc

Open clusters

NGC/IC	Other	Con	RA h m	Dec ° '	Mag	Diam '	N*	
4103		Cru	12 06.7	−61 15	7.4	7	45	
4349		Cru	12 24.5	−61 54	7.4	16	30	
4439		Cru	12 28.4	−60 06	8.4	4		
—	H5	Cru	12 29.0	−60 46	7.1	6		
4463		Mus	12 30.0	−64 48	7.2	5	30	
4609		Cru	12 42.3	−62 58	6.9	5	40	
4755		Cru	12 53.6	−60 20	4.2	10	50	Jewel Box, κ Crucis
4815		Mus	12 58.0	−64 57	8.6	3	100	
5138		Cen	13 27.3	−59 01	7.6	8	40	
5281		Cen	13 46.6	−62 54	5.9	5	40	
5316		Cen	13 53.9	−61 52	6.0	14	80	
5460		Cen	14 07.6	−48 19	5.6	25	40	
5606		Lup	14 27.8	−59 38	7.7	3	15	
5617		Cen	14 29.8	−60 43	6.3	10	80	
5662		Cen	14 35.2	−56 33	5.5	12	70	
5749		Lup	14 48.9	−54 31	8.8	8	30	
5822		Lup	15 05.2	−54 21	6.5	40	150	
5823		Cir	15 05.7	−55 36	7.9	10	100	
5925		Nor	15 27.7	−54 31	8.4	15	120	

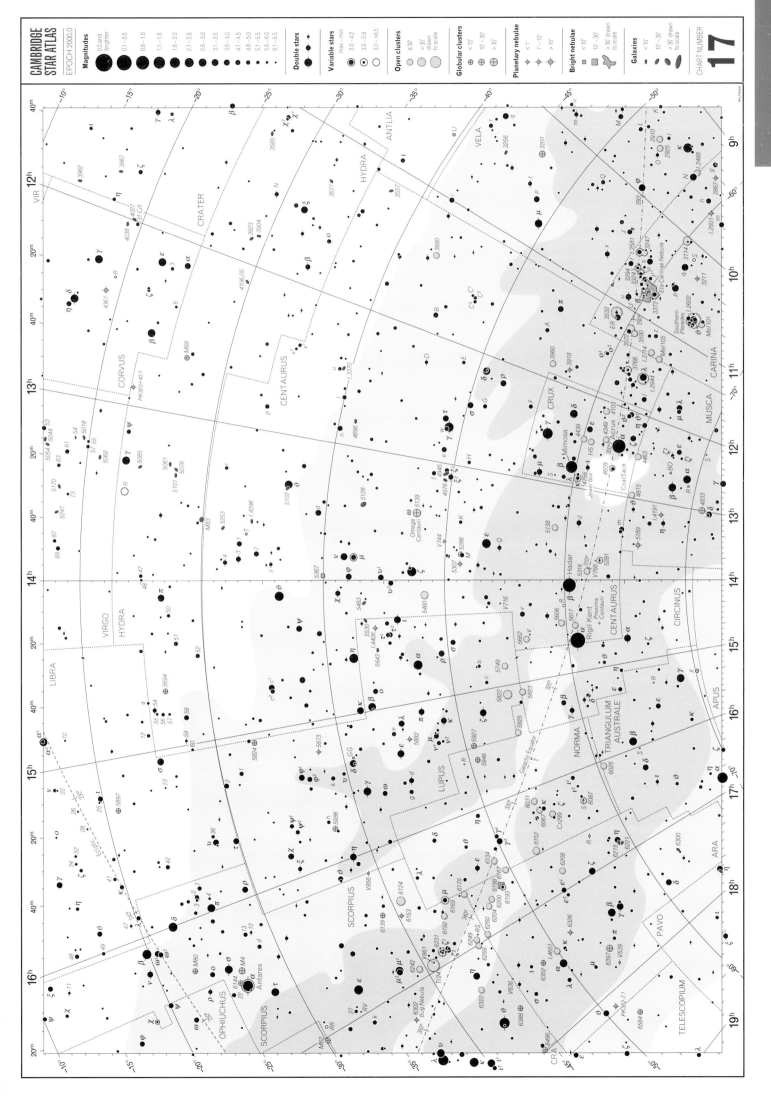

Chart 18 RA 16ʰ to 20ʰ, declination −20° to −65°

Variable stars

Star	Con	RA h m	Dec ° '	Type	Range	Period (days)	Spectrum
S	TrA	16 01.2	−63 47	Cep	6.1–6.8	6.32	F
S	Nor	16 18.9	−57 54	Cep	6.1–6.8	9.75	F–G
α	Sco	16 29.4	−26 26	SR	0.9–1.8	1733	M+B Antares
R	Ara	16 39.7	−57 00	EA	6.0–6.9	4.43	B
RS	Sco	16 55.6	−45 06	M	6.2–13.0	320.1	M
RR	Sco	16 56.6	−30 35	M	5.0–12.4	279.40	B
V861	Sco	16 56.6	−40 49	EB	6.1–6.7	7.85	B
RV	Sco	16 58.3	−33 37	Cep	6.6–7.5	6.06	F–G
V636	Sco	16 58.3	−45 37	Cep	6.0–6.9	6.80	G
BM	Sco	17 22.8	−32 13	SR	5.0–6.9	850:	K
X	Sgr	17 41.0	−27 50	Cep	4.2–4.8	7.01	F
V539	Ara	17 47.6	−53 37	EA	5.7–6.2	3.17	B+B
W	Sgr	17 50.5	−29 35	Cep	4.3–5.1	7.59	F–G
RS	Sgr	18 05.0	−34 06	EA	6.0–6.9	2.42	B
V1017	Sgr	18 17.6	−29 24	ZA?	6.2–14.7	—	G
λ	Pav	18 32.1	−62 11	Irr	3.4–4.3	—	B
RY	Sgr	18 52.2	−33 31	RCB	6.0–15.0	—	G
S	Pav	19 16.5	−59 12	SR	6.6–10.4	386.3	M
RR	Sgr	19 55.2	−29 11	M	5.6–14.0	334.6	M
RU	Sgr	19 55.9	−41 51	M	6.0–13.8	240.3	M

Double stars

Star	Con	RA h m	Dec ° '	PA °	Sep ''	Magnitudes	
12	Sco	16 12.3	−28 25	73	4.0	5.9+7.9	
σ	Sco	16 21.2	−25 36	273	20.0	2.9+8.5	
ε	Nor	16 27.2	−47 33	335	22.8	4.8+7.5	
α	Sco	16 29.4	−26 26	275	2.9	1.2v+5.4	Antares; var; bin., 878 y.
36	Oph	17 15.3	−26 36	146	4.9	5.1+5.1	Binary, 548.7 years
γ	Ara	17 25.4	−56 23	328	17.9	3.3+10.3	
η	Sgr	18 17.6	−36 46	105	3.6	3.2+7.8	
ξ	Pav	18 23.2	−61 30	154	3.3	4.4+8.6	
21	Sgr	18 25.3	−20 32	289	1.8	4.9+7.4	
λ	CrA	18 43.8	−38 19	214	29.2	5.1+9.7	
γ	CrA	19 06.4	−37 04	55	1.3	4.8+5.1	Binary, 120.4 years
π	Sgr	19 09.8	−21 01	150	0.1	3.7+3.7	
β¹	Sgr	19 22.6	−44 28	77	28.3	4.0+7.1	
52	Sgr	19 36.7	−24 53	170	2.5	4.7+9.2	

Open clusters

NGC/IC	Other	Con	RA h m	Dec ° '	Mag	Diam '	N*
6025		TrA	16 03.7	−60 30	5.1	12	60
6067	Cr 299	Nor	16 13.2	−54 13	5.6	13	100
—		Cep	16 18.4	−55 07	6.9	20	40
6087		Nor	16 18.9	−57 54	5.4	12	40
6124		Sco	16 25.6	−40 40	5.8	29	100
6134		Nor	16 27.7	−49 09	7.2	7	
6169		Lup	16 34.1	−44 03	6.6	7	40
6167		Nor	16 34.4	−49 36	6.7	8	
6178		Nor	16 35.7	−45 38	7.2	4	12
6193		Ara	16 41.3	−48 46	5.2	15	
6200		Ara	16 44.2	−47 29	7.4	12	40
6208		Ara	16 49.5	−53 49	7.2	16	60
6231		Sco	16 54.0	−41 48	2.6	15	
6242		Sco	16 55.6	−39 30	6.4	9	
6250		Ara	16 58.0	−45 48	5.9	8	60
6322		Sco	17 18.5	−42 57	6.0	10	30
I.4651		Ara	17 24.7	−49 57	6.9	12	80
6383		Sco	17 34.8	−32 34	5.5	5	40
6405	M6	Sco	17 40.1	−32 13	4.2	15	80
6416		Ara	17 44.4	−32 21	4.7	18	40
6425		Sgr	17 46.9	−31 32	7.2	8	35
6475	M7	Sco	17 53.9	−34 49	3.3	80	80
6520		Sgr	18 03.4	−27 54	7.6	6	60
6531	M21	Sgr	18 04.6	−22 30	5.9	13	17

Globular clusters

NGC/IC	Other	Con	RA h m	Dec ° '	Mag	Diam '
6093	M80	Sco	16 17.0	−22 59	7.2	8.9
6121	M4	Sco	16 23.6	−26 32	5.9	26.3
6144		Sco	16 27.3	−26 02	9.1	9.3
6139		Sco	16 27.7	−38 51	9.2	5.5
6266	M62	Oph	17 01.2	−30 07	6.6	14.1
6273	M19	Oph	17 02.6	−26 16	7.2	13.5
6284		Oph	17 04.5	−24 46	9.0	5.6
6287		Oph	17 05.2	−22 42	9.2	5.1
6293		Oph	17 10.2	−26 35	8.2	7.9
6304		Oph	17 14.5	−29 28	8.4	6.8
6316		Oph	17 16.6	−28 08	9.0	4.9
6352		Ara	17 25.5	−48 25	8.2	7.1
6388		Sco	17 36.3	−44 44	6.9	8.7
6397		Ara	17 40.7	−53 40	5.7	25.7
6441		Sco	17 50.2	−37 03	7.4	7.8
6496		Sco	17 59.0	−44 16	9.2	6.9
6544		Sgr	18 07.3	−25 00	8.3	8.9
6541		CrA	18 08.0	−43 42	6.7	13.1
6553		Sgr	18 09.3	−25 54	8.3	8.1
6569		Sgr	18 13.6	−31 50	8.7	5.8
6584		CrA	18 18.6	−52 13	9.2	7.9
6624		Sgr	18 23.7	−30 22	8.3	5.9
6626	M28	Sgr	18 24.5	−24 52	6.9	11.2
6638		Sgr	18 30.9	−25 30	9.2	5.0
6637	M69	Sgr	18 31.4	−32 21	7.7	7.1
6642		Sgr	18 31.9	−23 29	8.8	4.5
6652		Sgr	18 35.8	−32 59	8.9	3.5
6656	M22	Sgr	18 36.4	−23 54	5.1	24.0
6681	M70	Sgr	18 43.2	−32 18	8.1	7.8
6715	M54	Sgr	18 55.1	−30 29	7.7	9.1
6717		Sgr	18 55.1	−22 42	9.2	3.9
6723		Sgr	18 59.6	−36 38	7.3	11.0
6752		Pav	19 10.9	−59 59	5.4	20.4
6809	M55	Sgr	19 40.0	−30 58	7.0	19.0

Bright diffuse nebulae

NGC/IC	Other	Con	RA h m	Dec ° '	Type	Diam '	Mag*	
6188, 6193		Ara	16 40.5	−48 47	E+R	20×12		
6514	M20	Sgr	18 02.6	−23 02	E+R	29×27	7.6	Trifid Nebula
6523	M8	Sgr	18 03.8	−24 23	E	90×40	0.0	Lagoon Nebula

Planetary nebulae

NGC/IC	Other	Con	RA h m	Dec ° '	Mag	Diam "	Mag*	
6153		Sco	16 31.5	−40 15	11.5	25		
I.4634		Oph	17 01.6	−21 50	10.7	9	15 p	
6302		Sco	17 13.7	−37 06	12.8	50		Bug Nebula
6326		Ara	17 20.8	−51 45	12.2	14	13.5	
6369		Oph	17 29.3	−23 46	12.9	30/66	14.7	
—	PK352−7.1	Ara	18 00.2	−38 50	11.4	25		
I.4699		CrA	18 18.5	−45 59	11.9	10		
6629		Sgr	18 25.7	−23 12	11.6	15		
—	PK3−14.1	Sgr	18 55.6	−32 16	10.9	4		
I.1297		CrA	19 17.4	−39 37	10.7	7		

Galaxies

NGC/IC	Other	Con	RA h m	Dec ° '	Mag	Size	Type
6215		Ara	16 51.1	−58 59	11.8	2.0×1.6	Sc
6221		Ara	16 52.8	−59 13	11.5	3.2×2.3	SBc
6300		Ara	17 17.0	−62 49	11.1	5.4×3.5	SBc
I.4662		Pav	17 47.1	−64 38	11.4	2.2×1.4	Irr
6744		Pav	19 09.8	−63 51	9.0	15.5×10.2	SBb
6753		Pav	19 11.4	−57 03	11.9	2.5×2.2	Sb

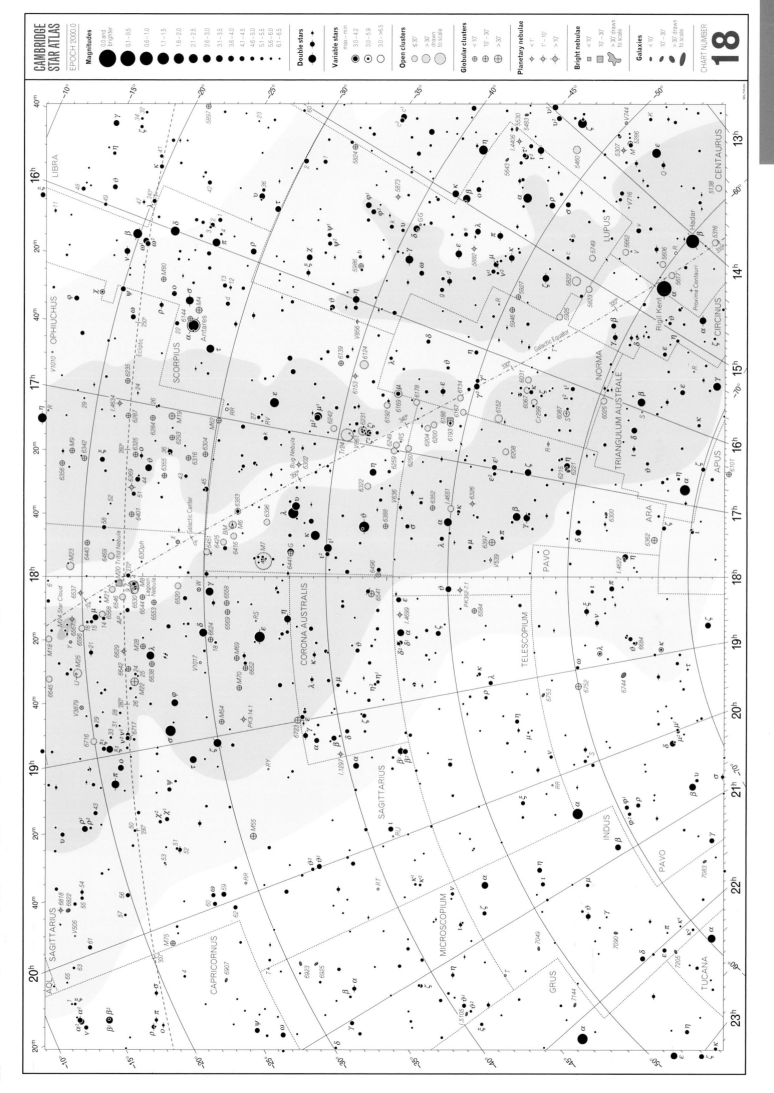

CAMBRIDGE STAR ATLAS

EPOCH 2000.0

CHART NUMBER

18

Chart 19   RA 20ʰ to 0ʰ, declination −20° to −65°

Variable stars

Star	Con	RA h m	Dec ° '	Range	Type	Period (days)	Spectrum
RR	Tel	20 04.2	−55 43	6.5-16.5 p	ZA	—	F
RT	Sgr	20 17.7	−39 07	6.0-14.1	M	305.3	M
T	Mic	20 27.9	−28 16	7.7-9.6 p	SR	344	M
T	Ind	21 20.2	−45 01	5.5-6.5	SR	320:	M
T	Gru	22 25.7	−37 34	7.8-12.3 p	M	136.5	M
S	Gru	22 26.1	−48 26	6.0-15.0	M	401.4	M

Double stars

Star	Con	RA h m	Dec ° '	PA °	Sep "	Magnitudes
α	Mic	20 50.0	−33 47	166	20.5	5.0 + 10.0
θ	Ind	21 19.9	−53 27	275	6.0	4.5 + 7.0
η	PsA	22 00.8	−28 27	115	1.7	5.8 + 6.8
41	Aqr	22 14.3	−21 04	114	5.0	5.6 + 7.1
δ	Tuc	22 27.3	−64 58	282	6.9	4.5 + 9.0
β	PsA	22 31.5	−32 21	172	30.3	4.9 + 7.9
γ	PsA	22 52.5	−32 53	262	4.2	4.5 + 8.0
δ	PsA	22 55.9	−32 32	244	5.0	4.2 + 9.2
υ	Gru	23 06.9	−38 54	211	1.1	5.7 + 8.0
θ	Gru	23 06.9	−43 31	75	1.1	4.5 + 7.0

Globular clusters

NGC/IC	Other	Con	RA h m	Dec ° '	Mag	Diam '
6864	M75	Sgr	20 06.1	−21 55	8.6	6.0
7099	M30	Cap	21 40.4	−23 11	7.5	11.0

Planetary nebula

NGC/IC	Other	Con	RA h m	Dec ° '	Mag p	Diam "	Mag*	
7293		Aqr	22 29.6	−20 48	7.3	769	13.8	Helix Nebula

Galaxies

NGC/IC	Other	Con	RA h m	Dec ° '	Mag	Size '	Type
6907		Cap	20 25.1	−24 49	11.3	3.4×3.0	SBb
6923		Mic	20 31.7	−30 50	12.1	2.5×1.4	Sb
6925		Mic	20 34.3	−31 59	11.3	4.1×1.6	Sb
7049		Ind	21 19.0	−48 34	10.7	2.8×2.2	S0
I5105		Mic	21 24.4	−40 37	11.5	2.5×1.5	E4
7083		Ind	21 35.7	−63 54	11.8	4.5×2.9	Sb
7090		Ind	21 36.5	−54 33	11.1	7.1×1.4	SBc
7144		Gru	21 52.7	−48 15	10.7	3.5×3.5	E0
7172		PsA	22 02.0	−31 52	11.9	2.2×1.3	S
7174		PsA	22 02.1	−31 59	12.6	1.3×0.7	S
7176		PsA	22 02.1	−31 59	11.9	1.3×1.3	E0
7184		Aqr	22 02.7	−20 49	12.0	5.8×1.8	Sb
7205		Ind	22 08.5	−57 25	11.4	4.3×2.2	Sb
7221		PsA	22 11.3	−30 37	12:	2.2×1.9	Sc
7314		PsA	22 35.8	−26 03	10.9	4.6×2.3	Sc
7410		Gru	22 55.0	−39 40	10.4	5.5×2.0	SBa
7412		Gru	22 55.8	−42 39	11.4	4.0×3.1	SBb
7418		Gru	22 56.6	−37 02	11.4	3.3×2.8	SBc
I1459		Gru	22 57.2	−36 28	10.0	3.4×2.5	E3
I5267		Gru	22 57.2	−43 24	10.5	5.0×4.1	S0
7424		Gru	22 57.3	−41 04	10.5	7.6×6.8	SBc
7456		Gru	23 02.1	−39 35	11.9	5.9×1.8	Sc
7496		Gru	23 09.8	−43 26	11.1	3.5×2.8	SBb
7531		Gru	23 14.8	−43 36	11.3	3.5×1.5	Sb
7552		Gru	23 16.2	−42 35	10.7	3.5×2.5	SBb
7582		Gru	23 18.4	−42 22	10.6	4.6×2.2	SBb
7599		Gru	23 19.3	−42 15	11.4	4.4×1.5	Sc
I5332		Scl	23 34.5	−36 06	10.6	6.6×5.1	Sd
7713		Scl	23 36.5	−37 56	11.6	4.3×2.0	SBd
7755		Scl	23 47.9	−30 31	11.8	3.7×3.0	SBb
7793		Scl	23 57.8	−32 35	9.1	9.1×6.6	Sd

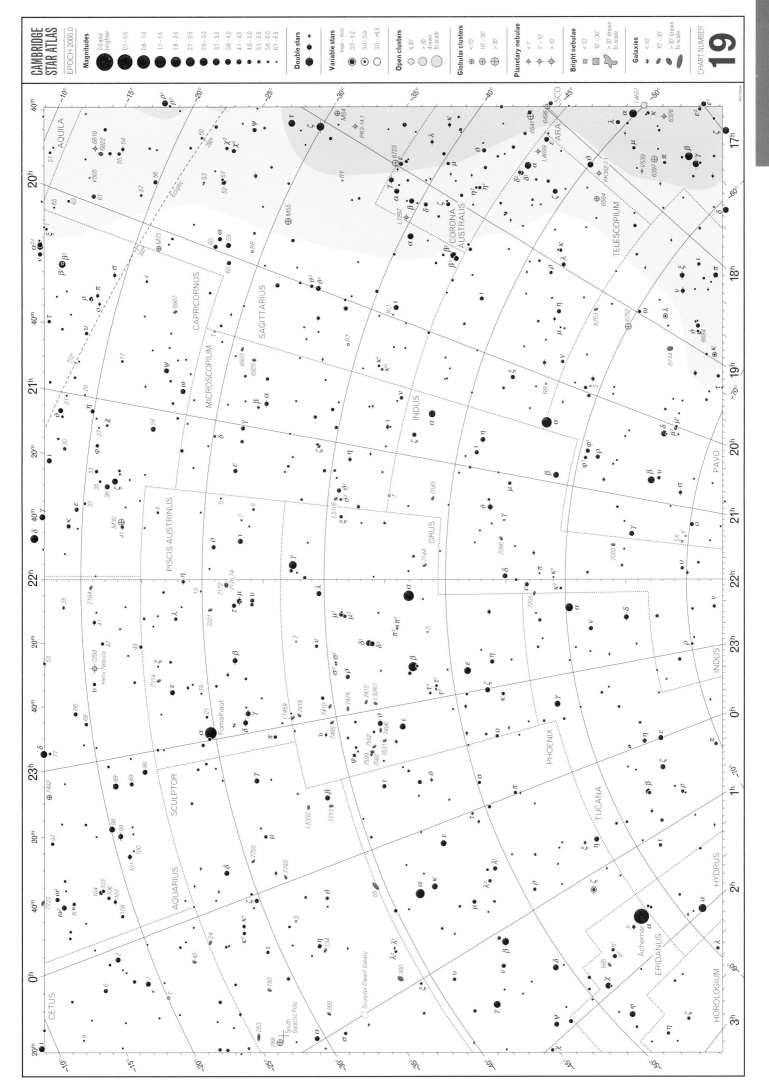

CAMBRIDGE
STAR ATLAS
EPOCH 2000.0

Magnitudes
0.0 and brighter
0.1–0.5
0.6–1.0
1.1–1.5
1.6–2.0
2.1–2.5
2.6–3.0
3.1–3.5
3.6–4.0
4.1–4.5
4.6–5.0
5.1–5.5
5.6–6.0
6.1–6.5

Double stars

Variable stars
max – min
3.0–4.2
3.0–5.9
3.0–>6.5

Open clusters
≤30'
>30'
drawn to scale

Globular clusters
<10'
10'–30'
>30'

Planetary nebulae
<1'
1'–10'
>10'

Bright nebulae
<10'
10'–30'
>30' drawn to scale

Galaxies
<10'
10'–30'
>30' drawn to scale

CHART NUMBER
19

W.Tirion

Chart 20 *South of declination −65°*

Variable stars

Star	Con	RA h m	Dec ° '	Range	Type	Period (days)	Spectrum
U	Men	04 09.6	−81 51	8.0–10.9 p	M	407	M
R	Oct	05 26.1	−86 23	6.4–13.2	M	405.6	M
TZ	Men	05 30.2	−84 47	6.2–6.9	EA	8.57	B
RS	Cha	08 43.2	−79 04	6.0–6.7	EA+δSct	1.67	A+A
S	Mus	12 12.8	−70 09	5.9–6.4	Cep	9.66	F
BO	Mus	12 34.9	−67 45	6.0–6.7	Irr	—	M
R	Mus	12 42.1	−69 24	5.9–6.7	Cep	7.48	B
J	Aps	13 08.1	−65 18	6.4–8.6	SR	199	B
X	TrA	15 14.3	−70 05	5.0–6.4	Irr	—	K
R	TrA	15 19.8	−66 30	6.4–6.9	Cep	3.39	G
κ	Pav	18 16.9	−67 14	3.9–4.6	Cep	9.09	F
Y	Pav	21 24.3	−69 44	5.6–7.3	SR	233.3	M
SX	Pav	21 28.7	−69 30	5.4–6.0	SR	50:	M

Double stars

Star	Con	RA h m	Dec ° '	PA °	Sep "	Magnitudes	
κ	Tuc	01 15.8	−68 53	336	5.4	5.1+7.3	
γ	Vol	07 08.8	−70 30	300	13.6	4.0+5.9	
ζ	Vol	07 41.8	−72 36	116	16.7	4.0+9.8	
κ	Vol	08 19.8	−71 31	57	65.0	5.4+5.7	
ε	Cha	11 59.6	−78 13	188	0.9	5.4+6.0	
β	Mus	12 46.3	−68 06	43	1.3	3.7+4.0	Binary, 383.1 years
ι	Oct	12 55.0	−85 07	230	0.6	6.0+6.5	
θ	Mus	13 08.1	−65 18	187	5.3	5.7+7.3	
$\delta^{1,2}$	Aps	16 20.3	−78 42	12	102.9	4.7+5.1	
$\mu^{1,2}$	Oct	20 41.7	−75 21	17	17.4	7.1+7.6	
λ	Oct	21 50.9	−82 43	70	3.1	5.4+7.7	

Open cluster

NGC/IC	Other	Con	RA h m	Dec ° '	Mag	Diam '	N*	
—	Mel 101	Car	10 42.1	−65 06	8.0:	14	50	Close to I.2602

Globular clusters

NGC/IC	Other	Con	RA h m	Dec ° '	Mag	Diam '	
104		Tuc	00 24.1	−72 05	4.0	30.9	47 Tucanae
362		Tuc	01 03.2	−70 51	6.6	12.9	
4372		Mus	12 25.8	−72 40	7.8	16.8	
4833		Mus	12 59.6	−70 53	7.4	13.5	
6101		Aps	16 25.8	−72 12	9.3	10.7	
6362		Ara	17 31.9	−67 03	8.3	10.7	

Bright diffuse nebula

NGC/IC	Other	Con	RA h m	Dec ° '	Type	Diam '	Mag*	
2070	30 Dor	Dor	05 38.7	−69 06	E	40×25		Tarantula Nebula in LMC[a]

Planetary nebulae

NGC/IC	Other	Con	RA h m	Dec ° '	Mag p	Diam "
I.2448		Car	09 07.1	−69 57	11.5	8
I.4191		Mus	13 08.8	−67 39	12.0	5
5189		Mus	13 33.5	−65 59	10.3	153

Galaxies

NGC/IC	Other	Con	RA h m	Dec ° '	Mag	Size '	Type	Mag*
1313		Ret	03 18.3	−66 30	9.4	8.5×6.6	SB	
2442		Vol	07 36.4	−69 32	11.2	6.0×5.5	SB	12.9
3059		Car	09 50.2	−73 55	12.0	3.2×3.0	SB	14.0
6684		Pav	18 49.0	−65 11	11.4	3.7×2.7	SB	

[a]LMC Large Magellanic Cloud: see page 83.

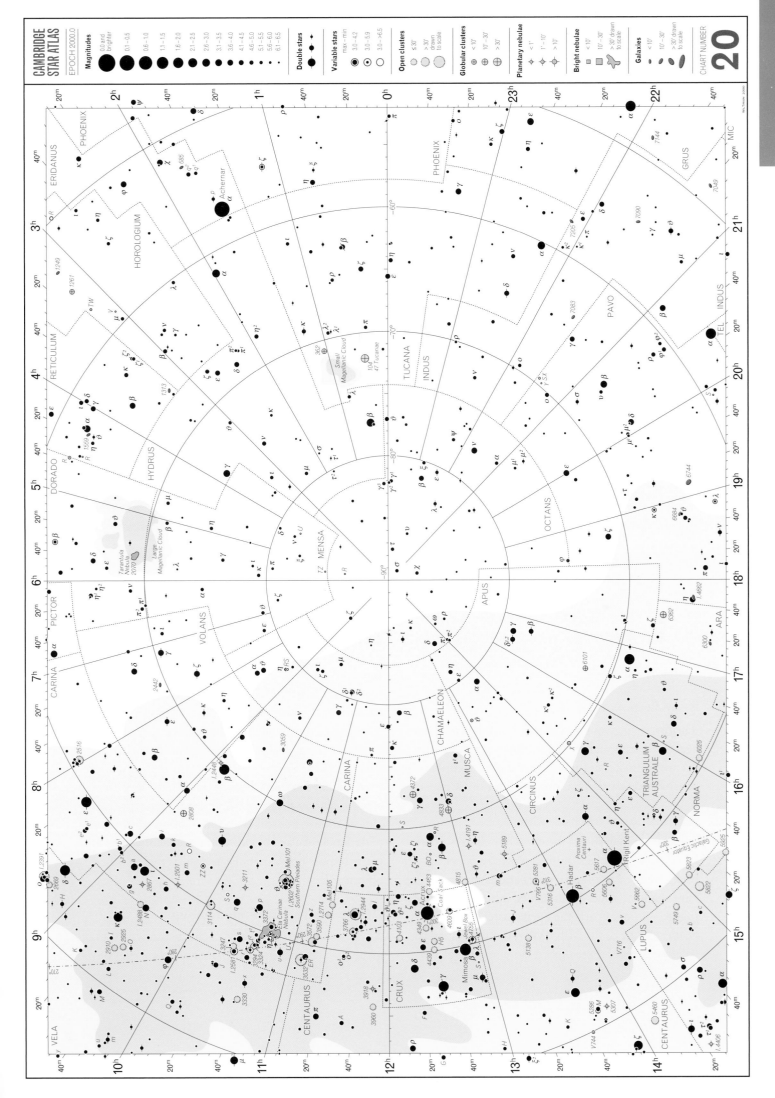

CAMBRIDGE STAR ATLAS

EPOCH 2000.0

Magnitudes

0.0 and brighter · 0.1–0.5 · 0.6–1.0 · 1.1–1.5 · 1.6–2.0 · 2.1–2.5 · 2.6–3.0 · 3.1–3.5 · 3.6–4.0 · 4.1–4.5 · 4.6–5.0 · 5.1–5.5 · 5.6–6.0 · 6.1–6.5

Double stars

Variable stars
max – min
3.0 – 4.2
3.0 – 5.9
3.0 – >6.5

Open clusters
≤30'
>30'
drawn to scale

Globular clusters
<10'
10'–30'
>30'

Planetary nebulae
<1'
1'–10'
>10'

Bright nebulae
<10'
10'–30'
>30' drawn to scale

Galaxies
<10'
10'–30'
>30' drawn to scale

CHART NUMBER

20

THE ALL-SKY MAPS

This final part of the *Cambridge Star Atlas* consists of six all-sky maps to show the general distribution of different objects in the sky. Each of the six maps shows the whole sky in a so-called 'equal-area' projection; the Mollweide projection, named after the man who invented it. 'Equal-area' means that in spite of the inevitable distortion the actual area covered by one square degree remains the same, no matter where on the map it is measured. So the distribution, or density of objects, is not influenced by the map's projection.

Unlike the main star charts, these maps show the galactic coordinates. The Galactic Equator is the central, horizontal line, marked 0°. On the main star charts it is represented as a line of dots and dashes, making an angle of almost 63° with the Celestial Equator. The Galactic Center (galactic longitude 0°, see star chart 18) lies close to the point where the constellations Sagittarius, Ophiuchus and Scorpius meet, and is in the center of the galactic maps in this chapter. At the top and bottom of the maps the North Galactic Pole (NGP) and the South Galactic Pole (SGP) are marked. They can be found on the star charts 5 (in Coma Berenices) and 14 (in Sculptor) respectively.

The constellations

The first map gives the positions of the constellations against this unusual grid. Stars down to magnitude 4.5 are plotted, plus some fainter stars to complete the constellation patterns, as on the monthly sky maps.

Distribution of open clusters

The second map shows the open star clusters as yellow disks with a black outline, plotted against the background of stars and constellations shown here in faint shades of blue. As explained in the introduction to the main star charts the open clusters are found near the plane of the Milky Way, so on this map you will find most of these objects close to the Galactic

Equator. The map shows all clusters that are plotted on the star charts. This is also the case with the globular clusters, the planetary nebulae and the galaxies.

Distribution of globular clusters

The third map shows the globular star clusters, using the same yellow symbols that are used on the star charts. You can see that the distribution is very different from the open clusters. The globulars form a huge halo around the Milky Way, so they are much more scattered than the open clusters. But since we are not in the center of our galaxy (in fact, we are closer to the edge than to the center) we see most globular clusters in the direction of the galactic heart (in the center of the map). In the opposite direction (near 180° galactic longitude, the left- and right-hand edges of the map) they are almost absent.

Distribution of diffuse nebulae

Because on the star charts of the *Cambridge Star Atlas* only a limited number of bright diffuse nebulae are drawn, additional nebulae have been added on the all-sky map, to show their distribution properly. The light green squares are the ones drawn in the *Cambridge Star Atlas*, while the darker green squares represent the additional nebulae, taken from *Sky Atlas 2000.0 (second edition)*. As you can see, the distribution of the nebulae is quite similar to that of the open star clusters. It brings us back to the disk of the galaxy. Like the open clusters, the diffuse nebulae are found along the spiral arms of the Milky Way. In some areas you see gaps, where the nebulae are almost absent, but this is only an illusion. Large areas of dark nebulae and clouds of dust block the light of many remote bright nebulae as well as that of other objects.

Distribution of planetary nebulae

The distribution of planetary nebulae (drawn as soft green disks with four spikes) does not show any similarity with that of the open or the globular clusters. They are not exclusively found in the spiral arms of the Milky Way, like the open star clusters, and not in a halo, like the globular clusters. They fill a disk-like area that is much thicker than the main Milky Way disk, in which the open clusters and the diffuse nebulae are found.

Distribution of galaxies

Since the galaxies, shown as red ovals, are not part of the Milky Way, but are in fact very distant 'Milky Ways' themselves, there is no relation between their distribution in the sky and the distribution of objects that belong to our own Milky Way. But, looking at the map, you will get the impression that there is a relationship, since other galaxies are almost absent in the area of the Milky Way close to the Galactic Equator. This, however, has nothing to do with their real distribution in space. The enormous amounts of gas and dust in our galaxy blocks the light of most of the distant galaxies seen in that direction. So they only *appear* to be absent. The galaxies are obviously grouped into what astronomers call 'clusters' and 'superclusters', with dimensions beyond our imagination.

In the bottom-right quadrant of the map you see two large galaxies. These are the Large and the Small Magellanic Clouds. The two clouds are regarded as satellites of our own Milky Way, and therefore are shown in blue on the star charts (14, 15 and 20). However, since they are officially catalogued as galaxies, they have been plotted on this map as well.

The constellations

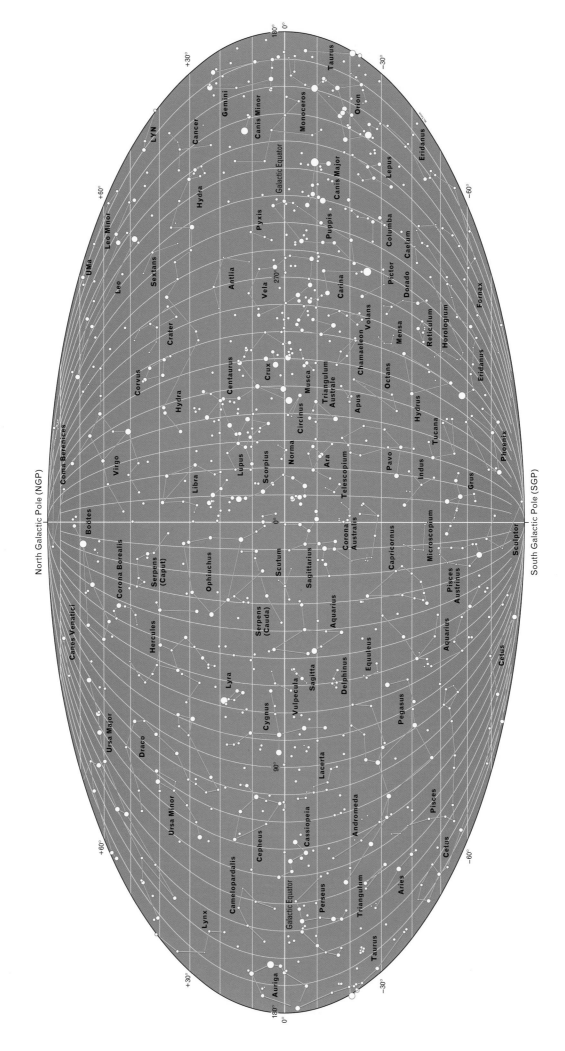

North Galactic Pole (NGP)

South Galactic Pole (SGP)

Mollweide's Equal-Area Projection

Galactic Coordinates

Distribution of open clusters

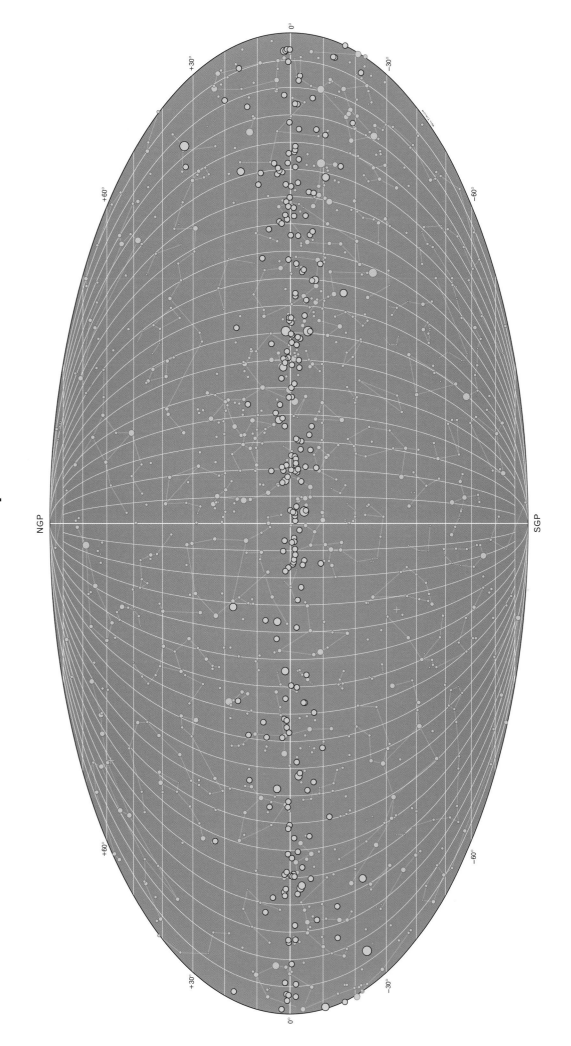

Mollweide's Equal-Area Projection

Galactic Coordinates

Distribution of globular clusters

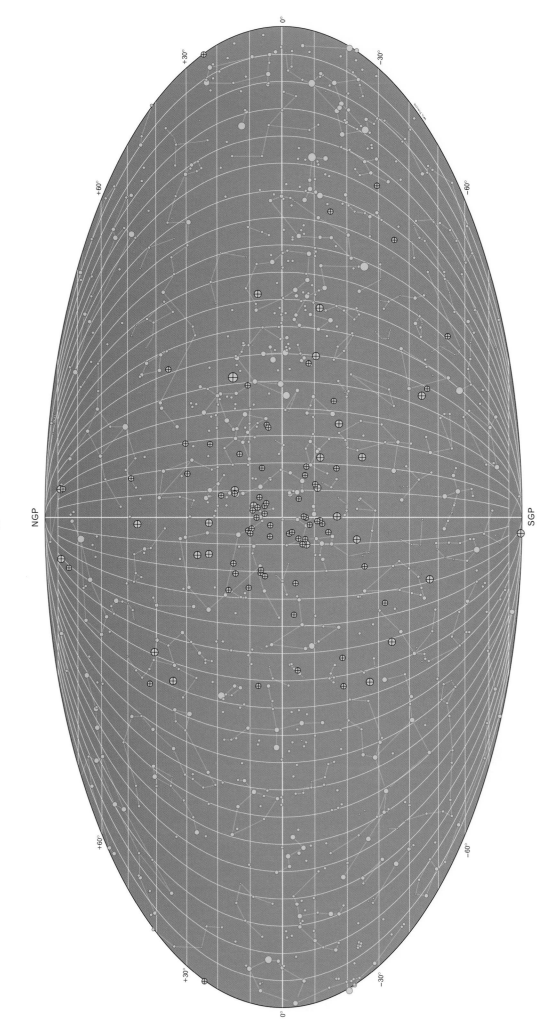

Mollweide's Equal-Area Projection

Galactic Coordinates

Distribution of diffuse nebulae

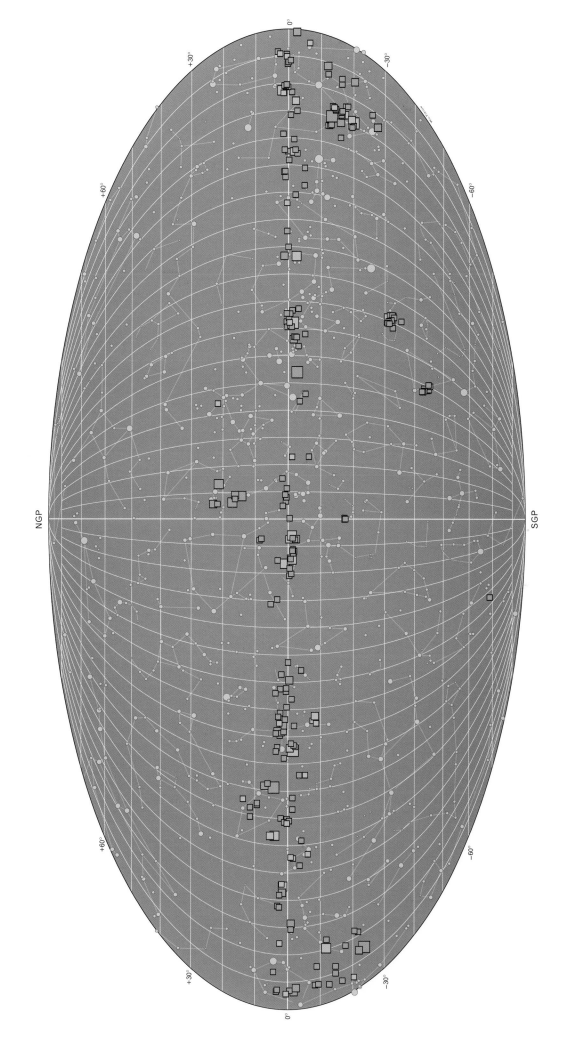

NGP

SGP

Mollweide's Equal-Area Projection

Galactic Coordinates

Distribution of planetary nebulae

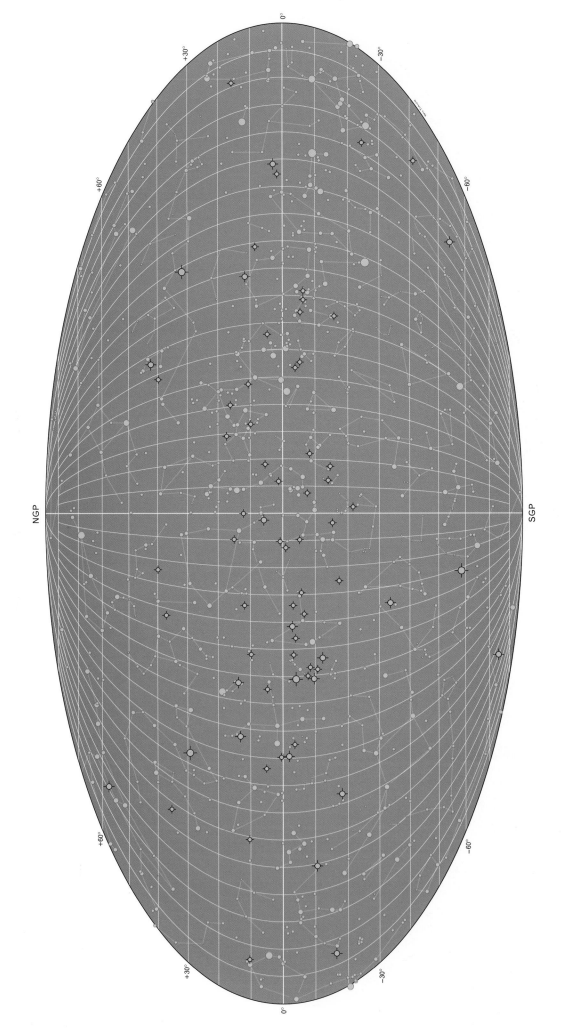

Mollweide's Equal-Area Projection

Galactic Coordinates

Distribution of galaxies

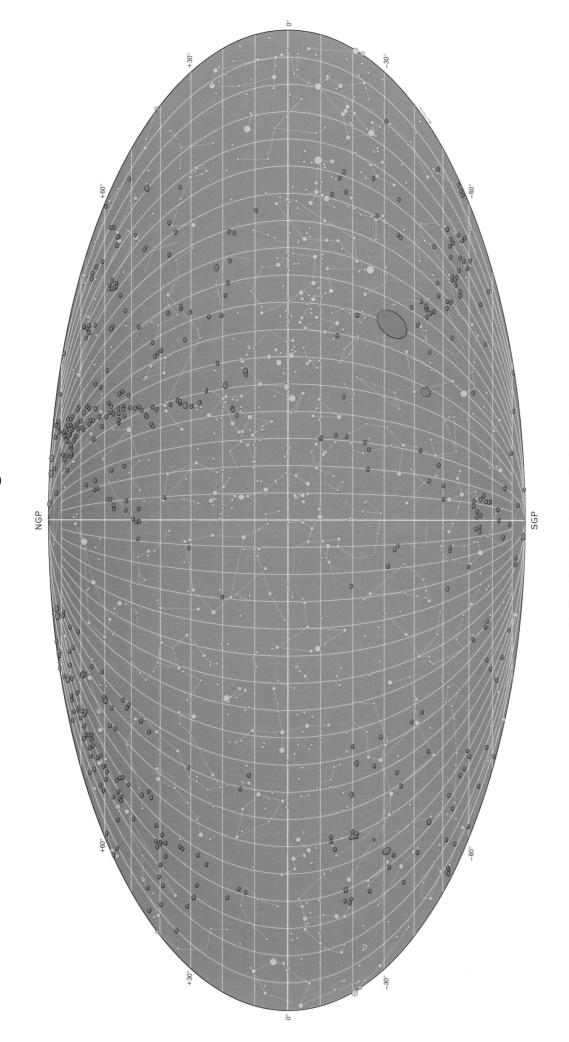

Mollweide's Equal-Area Projection

Galactic Coordinates

SOURCES AND REFERENCES

Catalogues

The Hipparcos and Tycho Catalogues
European Space Agency, ESTEC, Noordwijk,
The Netherlands, 1997

Sky Catalogue 2000.0, Volume 1, second edition
Alan Hirshfeld, Roger W. Sinnott and François
Ochsenbein (eds.)
Sky Publishing Corporation, Cambridge,
Massachusetts and Cambridge University Press,
Cambridge, England, 1991 (first edn 1982)

Sky Catalogue 2000.0, Volume 2
Alan Hirshfeld and Roger W. Sinnott (eds.)
Sky Publishing Corporation, Cambridge,
Massachusetts and Cambridge University Press,
Cambridge, England, 1985

NGC2000.0
Roger W. Sinnott (ed.)
Sky Publishing Corporation, Cambridge,
Massachusetts and Cambridge University Press,
Cambridge, England, 1988

Atlases

Millennium Star Atlas
Roger W. Sinnott and Michael A.C. Perryman
Sky Publishing Corporation, Cambridge,
Massachusetts,
European Space Agency, ESTEC, Noordwijk,
The Netherlands, 1997

Sky Atlas 2000.0, second edition
Wil Tirion and Roger W. Sinnott
Sky Publishing Corporation, Cambridge,
Massachusetts, 1998 (first edn 1981)
Cambridge University Press, Cambridge,
England, 1998 (first edn 1981)

Norton's Star Atlas, nineteenth edition
Ian Ridpath (ed.)
Longman Group UK Limited, 1998